Catholic Peacebuilding and Mining

Integral Peace, Development, and Ecology

Edited by Caesar A. Montevecchio and Gerard F. Powers

Routledge
Taylor & Francis Group
LONDON AND NEW YORK

First published 2022
by Routledge
2 Park Square, Milton Park, Abingdon, Oxon OX14 4RN

and by Routledge
605 Third Avenue, New York, NY 10158

Routledge is an imprint of the Taylor & Francis Group, an informa business

British Library Cataloguing-in-Publication Data
A catalogue record for this book is available from the British Library

Library of Congress Cataloging-in-Publication Data
A catalog record has been requested for this book

ISBN: 978-0-367-54508-6 (hbk)
ISBN: 978-0-367-55604-4 (pbk)
ISBN: 978-1-003-09427-2 (ebk)

DOI: 10.4324/9781003094272

Typeset in Sabon
by MPS Limited, Dehradun

Catholic Peacebuilding and Mining

This book explores the role of Catholic peacebuilding in addressing the global mining industry. Mining is intimately linked to issues of conflict, human rights, sustainable development, governance, and environmental justice. As an institution of significant scope and scale with a large network of actors at all levels and substantial theoretical and ethical resources, the Catholic Church is well positioned to acknowledge the essential role of mining, while challenging unethical and harmful practices, and promoting integral peace, development, and ecology. Drawing together theology, ethics, and praxis, the volume reflects the diversity of Catholic action on mining and the importance of an integrated approach. It includes contributions by an international and interdisciplinary range of scholars and practitioners. They examine Catholic action on mining in El Salvador, Peru, Colombia, the Democratic Republic of the Congo, and the Philippines. They also address general issues of corporate social responsibility, human rights, development, ecology, and peacebuilding. The book will be of interest to scholars of theology, social ethics, and Catholic studies as well as those specializing in development, ecology, human rights, and peace studies.

Caesar A. Montevecchio is Assistant Director of the Catholic Peacebuilding Network based in the Kroc Institute for International Peace Studies of the Keough School of Global Affairs at the University of Notre Dame, USA.

Gerard F. Powers is Coordinator of the Catholic Peacebuilding Network and Director of Catholic Peacebuilding Studies in the Kroc Institute for International Peace Studies of the Keough School of Global Affairs at the University of Notre Dame, USA.

Routledge Studies in Religion

Former Muslims in Europe
Between Secularity and Belonging
Maria Vliek

Religion and Violence in Western Traditions
Selected Studies
Edited by André Gagné, Jennifer Guyver, and Gerbern S. Oegema

Jewish Approaches to Hinduism
A History of Ideas from Judah Ha-Levi to Jacob Sapir (12th–19th centuries)
Richard G. Marks

Spiritual, Religious, and Faith-Based Practices in Chronicity
An Exploration of Mental Wellness in Global Context
Edited by Andrew R. Hatala and Kerstin Roger

An Anthropology of the Qur'an
Ahmed Achrati

Contrasts in Religion, Community, and Structure at Three Homeless Shelters
Changing Lives
Ines W. Jindra, Michael Jindra and Sarah DeGenero

Enhancement Fit for Humanity
Perspectives on Emerging Technologies
Michael Baggot

Catholic Peacebuilding and Mining
Integral Peace, Development, and Ecology
Edited by Caesar A. Montevecchio and Gerard F. Powers

Religion, Discourse, and Society
Towards a Discursive Sociology of Religion
Marcus Moberg

For more information about this series, please visit: https://www.routledge.com/religion/series/SE0669

To the memory of Robert J. Schreiter, C.PP.S. (1947–2021), an inspiring artisan of peace.

Contents

Acknowledgments

This book is an initiative of the Catholic Peacebuilding Network (CPN). It would not have been possible without the sustained and generous support of CPN's secretariat by the University of Notre Dame's Kroc Institute for International Peace Studies, part of the Keough School of Global Affairs. We are grateful to the Kroc Institute for hosting the authors' colloquium, and to Caritas Internationalis, Catholic Relief Services, the Catholic Theological Society of America, the Catholic University of America's Institute for Policy Research, Georgetown University's Berkley Center for Religion, Peace and World Affairs, the US Conference of Catholic Bishops' Office of International Justice and Peace, and the University of Notre Dame's Pulte Institute for Global Development of the Keough School for hosting events that contributed to this endeavor. Since the initiative was conceived in 2014, the CPN Steering Committee has provided invaluable support and guidance. We are especially grateful to the contributors for their patience and dedication to the project, especially given the limitations and stresses of the COVID-19 pandemic. And we thank Catherine R. Osborne for providing copy editing services.

Contributors

Albino Barrera, OP, is a professor of theology and economics at Providence College. He is a friar of the St. Joseph Province (New York) of the Order of Preachers. His most recent book is *Catholic Missionaries and Their Work with the Poor: Mitigating Market-Government Failure in Emerging Nations* (2019).

Douglass Cassel is professor emeritus in the Law School at the University of Notre Dame and a Notre Dame Presidential Fellow. He is a scholar, practitioner, and commentator on international human rights law, specializing in issues of business and human rights, regional human rights systems, and international criminal and humanitarian law.

Karl M. Gaspar, CSsR, is a professor of missiology at the St. Alphonsus Theological and Mission Institute (Davao City), and a Redemptorist brother. He serves as a consultant to the Philippine Bishops' Commission on Indigenous Peoples and has been engaged in multiple social initiatives promoting indigenous rights, human rights, peacebuilding, and interfaith dialogue.

Héctor Fabio Henao is a Catholic priest of the Archdiocese of Bogotá in Colombia. He is the director of Cáritas Colombiana and of the National Social Pastoral of the Episcopal Conference of Colombia. He is a past president of Colombia's National Council for Peace, Reconciliation, and Coexistence, and currently is coordinator of its Territorial Peace Commission.

William N. Holden is an associate professor in the Department of Geography and B.Sc. in Environmental Science at the University of Calgary. He is also a research associate with the Canadian Institute of Resources Law. His current work focuses on the Philippines, climate change, the efficacy of mining as a development strategy, insurgency/counterinsurgency warfare, state terrorism, and liberation theology and Maoism as counter hegemonic discourses in the twenty-first century.

Laurie Johnston is an associate professor of theology and religious studies at Emmanuel College (Boston). Her specialty is Christian social ethics, particularly questions of war and peacemaking. She is the co-editor of *Can War Be Just in the 21st Century? Ethicists Engage the Tradition* (2015, with Tobias Winright) and *Public Theology and the Global Common Good: the Contribution of David Hollenbach* (2016, with Kevin Ahern, Meghan Clark, and Kristin Heyer).

Katherine Marshall is a senior fellow at Georgetown University's Berkley Center for Religion, Peace, and World Affairs, where she leads the center's work on religion and global development. She is also a professor of the practice of development, conflict, and religion in Georgetown's Walsh School of Foreign Service. She helped create and now serves as the executive director of the World Faiths Development Dialogue. She is the co-editor of *Women, Religion, and Peacebuilding: Illuminating the Unseen* (2015, with Susan Hayward).

Andrés McKinley is a water and mining specialist at the José Simeón Cañas Central American University (San Salvador). He has worked for 46 years with international organizations on sustainable development and social justice. His most recent book is *La Crisis Hidrica en El Salvador: Hechos y Retos* (*The Water Crisis in El Salvador: Facts and Challenges*) (2018).

Vincent J. Miller is the Gudorf Chair of Catholic Theology and Culture at the University of Dayton. He has authored numerous articles and book chapters on ecology, theology and culture, and theology and public life. He edited *The Theological and Ecological Vision of Laudato Si': Everything is Connected* (2017) and wrote *Consuming Religion: Christian Faith and Practice in a Consumer Culture* (2005).

Rigobert Minani, SJ, is a Congolese Jesuit. He is the head of research for the Peace, Human Rights, Democracy, and Good Governance Department at the Centre d'Etude Pour l'Action Sociale in the DRC. He is also team leader of the Ecclesial Network of the Congo Basin Forest and director of Jesuit Social Ministry in DRC and Angola. He is the editor of *Unmet Promises of Extractive Industries in Africa* (2017).

Caesar A. Montevecchio is assistant director of the Catholic Peacebuilding Network at the University of Notre Dame's Kroc Institute for International Peace Studies, Keough School of Global Affairs. He coordinates the Catholic Peacebuilding Network's research initiative on peacebuilding and mining. He is the co-author of *Catholic Approaches to Transitional Justice and Reconciliation: Guidelines for Reflection and Planning* (2018, with Tom Bamat and Gerard F. Powers).

Raymond Offenheiser leads the University of Notre Dame's Pulte Institute for Global Development, Keough School of Global Affairs. He is also a distinguished professor of the practice at the Keough School. Prior to joining Notre Dame, Offenheiser was the president of Oxfam America, where his work included extensive engagement on issues of mining, development, and justice.

Elias O. Opongo, SJ, is senior lecturer of peace studies at the Hekima Institute of Peace Studies and International Relations, Hekima University College (Nairobi), and director of the Centre for Research, Training, and Publications at the same university. His research interests include transitional justice, post-conflict reconstruction, extractive industries and conflict, state-building and democracy, and religious extremism and violence. His books include *Extractive Industries in Africa: Addressing Conflicts and Integrating Sustainable Development* (2018) and *Informal Mining and Family Vulnerability in Kenya, Zambia, and Zimbabwe* (2016).

José Bayardo Chata Pacoricona is a lawyer working with the Human Rights, Indigenous Rights, and Environment section of Derechos Humanos y Medio Ambiente—Puno (DHUMA). DHUMA began as the Vicariate of Solidarity within the Catholic Prelature of Juli, Peru. It is now an independent NGO that focuses on defending the legal rights of indigenous peoples and the environment in the Puno region of Peru.

Sandra Polanía-Reyes is an associate professor of economics at the University of Navarra (Pamplona). She is a development economist working at the intersection of experimental economics and prosocial behavior. Her research focuses on the design and evaluation of effective behavioral policy interventions in developing countries.

Gerard F. Powers is director of Catholic Peacebuilding Studies and coordinator of the Catholic Peacebuilding Network at the Kroc Institute for International Peace Studies, Keough School of Global Affairs, University of Notre Dame. He specializes in the ethics of the use of force; religion, conflict, and peacebuilding; and ethics and U.S. foreign policy. He is the co-editor of *Peacebuilding: Catholic Theology, Ethics and Praxis* (2010, with Robert J. Schreiter and R. Scott Appleby) and *Strategies of Peace: Transforming Conflict in a Violent World* (2010, with Daniel Philpott).

Anna Floerke Scheid is an associate professor of theology at Duquesne University (Pittsburgh). Her research focuses on the ethics of war and peacebuilding, restorative justice, political resistance, religion and violence, and racial justice. She is the author of *Just Revolution: A Christian Ethic of Political Resistance and Social Transformation* (2015).

Daniel P. Scheid is an associate professor of theology at Duquesne University (Pittsburgh). His work focuses on interreligious ecological ethics, drawing on Catholic social thought and comparative theology, with particular attention to Catholic-Hindu and Catholic-Buddhist dialogue.

He is the author of *The Cosmic Common Good: Religious Grounds for Ecological Ethics* (2016).

Clemens Sedmak is a professor of social ethics in the Keough School of Global Affairs at the University of Notre Dame. He also is a concurrent professor at Notre Dame's Center for Social Concerns and in its Department of Theology, and is interim director of the Nanovic Institute for European Studies. His books include *The Capacity to be Displaced: Resilience, Mission, and Inner Strength* (2017) and *A Church of the Poor: Pope Francis and the Transformation of Orthodoxy* (2016).

Cardinal Peter K.A. Turkson is prefect of the Vatican's Dicastery for Promoting Integral Human Development, a position he has held since the Dicastery's inception in 2017. Previously he was president of the Pontifical Council for Justice and Peace (2009–2017) and archbishop of Cape Coast in Ghana (1992–2009). He has been a chancellor of the Catholic University College of Ghana since 2003.

Tobias Winright is an associate professor in both the Department of Theological Studies and the Albert Gnaegi Center for Health Care Ethics at Saint Louis University. His research and teaching interests include just war theory, criminal justice ethics, ecological theology, Catholic social thought, and bioethics. His publications include *Green Discipleship: Catholic Theological Ethics and the Environment* (2011) and *Can War Be Just in the 21st Century? Ethicists Engage the Tradition* (2015, co-editor with Laurie Johnston).

Introduction

Caesar A. Montevecchio and Gerard F. Powers

Mining is essential to life as we know it. Power plants, solar arrays, cell phones, computers, buildings, cars, and an infinite number of other items depend on mining. Yet mining is also a factor in many of the world's most intractable conflicts, intimately linked to violence, human rights abuses, environmental degradation, unsustainable development practices, and poor governance. Whether mining is a force for good or ill depends on many factors: the material mined, end uses and users, policies and practices of mining companies, the quality of governance, the role of affected communities, socio-economic-cultural contexts, and mining's place in broader development plans. As many authors in this book point out, there is no one-size-fits-all ethical approach to mining.

The relevance of mining to Catholic peacebuilding has been evident as the Catholic Peacebuilding Network (CPN),[1] a network of two dozen church institutions, universities, development agencies, and peace organizations, has accompanied the church in the Philippines, Colombia, and the Great Lakes Region of Africa. As this book details, the taxonomy of Catholic engagement on mining is varied. In the Democratic Republic of the Congo (DRC), where Rigobert Minani claims mining is "a structural cause of conflict," the church has led efforts to promote transparency in the mining industry, including working with an international coalition in support of the Dodd-Frank legislation and playing an official role in reviewing dozens of government mining contracts. Other examples include the Philippines, where the church opposed President Rodrigo Duterte's 2021 decision to lift a moratorium on mining contracts in key areas, and El Salvador, where the church played a key role in that country becoming the first nation to ban metal mining in 2017. In Colombia, church engagement primarily involves pastoral accompaniment and advocacy on behalf of local communities impacted by mines, and facilitating dialogue among communities, government, and corporations. And in Peru, Derechos Humanos y Medio Ambiente—Puno takes legal action to protect the human rights of rural and indigenous communities.

Reflecting this diversity of Catholic action on mining, this book argues for an integrated approach to mining that has three dimensions. First, to be

DOI: 10.4324/9781003094272-101

conceptually adequate, it must marry a Catholic theology and ethics of peace, justice, human rights, development, and ecology; all are connected, none alone is sufficient. Second, it needs to connect ethics and theological reflection with church practice on mining, which would strengthen both. Third, church engagement on mining should integrate local, national, and international dimensions (vertical integration), as well as diverse types of engagement by different sectors within the Catholic community (horizontal integration).

An integrated approach to Catholic peacebuilding and mining

The need for an integrated approach to mining is clear. From a global perspective, Katherine Marshall points out that mining is a factor in each of the seventeen UN Sustainable Development Goals, from ending poverty and ensuring clean water to promoting food security and peace. This interconnectedness is evident in Colombia. The government, supported by key international actors, is counting on a dramatic expansion of mining—much of it in politically-unstable and environmentally-sensitive areas of the Amazon formerly controlled by rebels from the Revolutionary Armed Forces of Colombia —to promote economic development and help deliver a peace dividend after the 2016 peace accord. Sandra Polonía-Reyes and Héctor Fabio Henao point out some of the difficulties in doing so. They conclude that, while there is "no linear relation between conflict and mining, ... mining intersects and interacts with violent conflict in numerous regions" of the country. In some cases, armed groups rely on mining to fund violent activity; in others, government or private security forces use violence against those opposing a mine; in still others, a mine might generate conflict within a community among those who see the mine as essential for development and those who are concerned about its negative impacts on the environment or indigenous rights. Despite the connection between mining and conflict, the fact that the 2016 accord, the world's most comprehensive, did not address mining was "a missed opportunity to connect integral peace, ecology, and development."

This book addresses this first dimension of integration by answering Pope Francis' (2019) call for the Catholic community to examine mining through the lens of his teaching on integral ecology, integral human development, and integral peace in *Laudato Si'*. Considerable scholarly work has been done on *Laudato Si'*, as well as Catholic perspectives on integral human development (see Cichos et al. 2021; Carozza and Sedmak 2020; Ilo 2014), ecology (see DiLeo 2017; Scheid 2016), corporate social responsibility and business ethics (see Finn 2021; Rebman 2020), and peacebuilding (see Hawksley 2020; Schreiter 2010), in general. But relatively little has been done to relate these topics to each other and to apply them, in an integrated way, to mining, in particular.[2]

Several chapters of this book take up the theoretical challenges of integrating these different dimensions of Catholic teaching. Others address more practical challenges, such as the need to develop the business, technical, legal, and financial sophistication on mining necessary for the church to be credible and effective. Three church responses to these challenges are common: mining is not a church issue, mining is good because it creates jobs, or the local community should decide. The first reflects a narrow ecclesiological response: the church is not competent to address mining because it lacks the secular expertise. The second focuses narrowly on just one aspect of integral development. The third prioritizes process over substance, and the local common good, often without adequate consideration of the national and global common good. Each response is legitimate, but the claim of this book is that they are ultimately inadequate because they fail to take the integral approach called for by Catholic social teaching.

A second dimension of integration involves practical theology, connecting theology and ethics to church practices. Practical theology, according to Robert Schreiter (2010, 366), "is an ongoing practice of reflection and action that keeps theory and informed practice in constant conversation with each other." The Catholic community is deeply engaged on mining in many places[3] but, unfortunately, this aspect of lived Catholicism has gone largely unexamined by and is not deeply grounded in Catholic theology and ethics.[4] Theological and ethical reflection must catch up to Catholic praxis, learn from it, and help develop new frameworks and ideas to strengthen and support it, consistent with *Laudato Si*'s integral approach. By putting scholars and church actors in conversation with each other, this book is a modest attempt to fill this gap.

A third dimension of integration—horizontal and vertical—relates to why a book would focus specifically on Catholic approaches to mining. Mining is simultaneously radically local, in how intimately operations are connected to specific places, and massively global, in how the industry is part of complex transnational economic networks and is dominated by relatively few multi-national companies. The Catholic Church is one of the few institutions in the world with the scope, scale, and sophistication needed to match the scope, scale, and sophistication of mining industries. It has a rich tradition of reflection on peace, development, human rights, and ecology necessary to address the many dimensions of mining. It has an institutional capacity to address a range of problems from the local to the global levels, and across national divides.[5] And its people power includes respected moral leaders at all levels as well as the capacity to catalyze a large network of grassroots actors.

This book highlights the distinct contributions that the Catholic Church, as a transnational institution, can make, as well as the challenges the church faces in using its capacity for horizontal and vertical integration. While the church is doctrinally hierarchical, it operates in a highly decentralized way.

It is not surprising, therefore, that much Catholic action is similar to that in Colombia: a mine-by-mine approach in which each case is addressed (or not) at the local level, often relying on some form of free, prior, and informed consent as a framework for pastoral engagement. In some countries, as is discussed in the chapters on the DRC, the Philippines, and El Salvador, the church does not engage only at the local level but has developed national policies on mining. While regional and continental episcopal conferences have occasionally addressed mining issues, in the past decade, the church has recognized the need for more focused regional responses, developing new initiatives to facilitate coordination of church actors in two of the world's most sensitive areas, the Amazon and Congo Basin. At the global level, the Holy See has convened consultations with church actors as well as with the top executives of the world's largest mining companies. The work being done at these different levels and transnationally is impressive. But, as this book underscores, the challenge of translating the capacity for an integrated approach into actual collaboration among these diverse church actors at different levels is daunting.

The structure of this book

The book is divided into two primary sections, with one lead chapter. Any reflection on ethics in Catholic tradition is moored in sacred Scripture. The first chapter by Cardinal Peter Turkson, Prefect of the Vatican's Dicastery for Promoting Integral Human Development, provides a scriptural foundation for an ethics of mining and peace. Cardinal Turkson examines the way mining is present in texts about God's covenant with Israel, in the wisdom tradition as reflected in the Book of Job, and in the perfection of the covenant in Christ. Situating mining in these contexts offers a challenge to the mining industry to recognize the ethical values associated with the kingdom of God and provides a foundation for incorporating mining into areas of theological ethics, like ecology, human rights, and peace.

The first section is about mining and lived religion. It begins with a framing essay by Katherine Marshall on the role of a diverse set of religious actors in advocacy and action on mining. As a corrective to the fact that religious actors do not have an accepted place in mining debates, she calls for "a well-conceived, collaborative, multireligious engagement" on ethical challenges associated with mining. Her chapter is followed by essays on Catholic engagement on mining in specific countries. With support from CPN, Caritas Colombia developed a database to track diocesan-level mining engagement in the country. Based on that data, Sandra Polanía-Reyes and Msgr. Héctor Fabio Henao propose a pastoral framework for expanded and more coordinated engagement by the church at the local, national, and international levels, especially for the sake of supporting Colombia's 2016 peace agreement. Rigobert Minani, SJ, then addresses the DRC and some of the most systematic Catholic efforts anywhere to address mining and conflict.

He focuses especially on the church's efforts to promote transparency and good governance, and to use the church's growing expertise on mining and Catholic social teaching to help reveal the limits of the promises of the extractive industry. Karl M. Gaspar, CSsR, traces the way mining gradually became a focal point of the peacebuilding work of Catholic leaders in the Philippines and explains how mining remains closely associated with violations against indigenous peoples and continuing violence, especially in the southern region of Mindanao. In Peru, the organization Derechos Humanos y Medio Ambiente—Puno (DHUMA) does extensive legal work to defend indigenous communities against mining. José Bayardo Chata Pacoricona outlines some of the major legislative and legal challenges DHUMA has faced. Finally, Andrés McKinley, who worked closely with Catholic leaders in El Salvador on the world's first national ban on metal mining, explains how the country's water crisis motivated El Salvador's people and Catholic leadership to mobilize for the ban.

The second section considers these and other cases in light of broader themes in Catholic theology and ethics. Tobias Winright, analogizing to the just war tradition, proposes an ethics of extraction that includes a just mining theory with three categories: *jus ad extractionem*, *jus in extractionem*, and *jus post extractionem*. Anna Floerke Scheid and Daniel P. Scheid undertake a constructive project to craft a moral framework to support mining and peace. They do so by combining the idea of integral ecology with the principles and practices of just peace theory, yielding a model of "ecological just peace."

In the following chapters, Douglass Cassel, Clemens Sedmak, Albino Barrera, OP, and Elias O. Opongo, SJ, give insights on mining in light of dialogue between Catholic and secular ideas on human rights, development, and governance. Cassel suggests that international human rights and Catholic social teaching on mining are complementary. International law provides the Catholic ethical vision with a practical vehicle for change, while the Catholic vision offers an ethical horizon that goes beyond legal norms and can drive more substantial change. Sedmak examines how the theological underpinnings of integral human development contrast with typical models of development in the mining sector in that peace and human flourishing are more important goals than economic growth. This means that the animating goal of profit maximization that typically drives mining must be restrained. In a similar vein, Barrera passes mining through the filter of development ethics and argues that the inherent limitations of the market prevent it from being sufficient to regulate the problems of injustice and conflict that can arise from mining. Finally, Opongo, citing the role of corruption and weak government regulation in mining-related conflicts in Africa, proposes that the Catholic community do more to leverage its strengths to promote good governance.

The remaining chapters from Vincent J. Miller, Raymond Offenheiser, and William N. Holden and Caesar A. Montevecchio consider the globalized context of mining and the fraught dynamics between the Global North and Global South. Miller argues that the problems with mining are a network problem more than a failure of moral will. While Catholics are united in Christ, church networks of communion are no match for mining market networks. Therefore, he recommends a "synodal ecclesial communications network" to deepen relationships among Catholic consumers and Catholics impacted by mining. Offenheiser builds on his experience engaging mining industry leaders and urges Catholic actors to take a nuanced approach that acknowledges progress the mining industry has made in addressing ethical issues in recent years. The Catholic community needs to better understand the industry's operations, priorities, and concerns in order to engage it successfully and bring about even more structural improvement for peace and the common good. Hardrock mining for metals and minerals is often forgotten in analyses of climate change and mining, but Holden and Montevecchio present four ways in which the interfaces of climate change and hardrock mining impact justice and peace: by contributing to climate injustice; by entrenching an "extractivist" model of economic development; by engendering violence against environmental and human rights activists; and through increased demand for uranium for nuclear power which increases risks of nuclear proliferation.

In the book's conclusion, Laurie Johnston identifies some strengths in the Catholic approach to mining and peace as well as some challenges and areas of needed growth: a focus on right relationships, but a need to improve the church's institutional transparency; cooperation with robust partnership networks, but a need to better include women's experiences and leadership; a focus on a big-picture common good, but a need to accept difficult tradeoffs; a capacity for prophetic criticism, but a need to balance it with practical cooperation; and consideration of the long-term, but a need to re-think aspects of Catholic social thought that reflect lingering anthropocentrism.

Aspirations of the book

This book is part of a wider effort by CPN to address issues of mining and peace. Prior to embarking on this book, CPN convened three colloquia for scholars and practitioners to explore the state of the question and the need for a project on mining. An additional colloquium was held for authors of this book. To complement these colloquia, CPN convened a three-year interest group on mining and peace under the auspices of the Catholic Theological Society of America. As noted, CPN also collaborated with Caritas Colombia on an extensive database on Catholic engagement on extractives. In addition to these efforts to improve understanding of the issue, CPN has served on the conflict resources and peacebuilding working group of Caritas Internationalis and has participated in several of the Holy See's meetings with mining executives and Catholic actors mentioned above.

Given that this book arises out of CPN's wider efforts to connect scholars and church actors engaging mining, it has two main audiences: scholars and practitioners. Theologians, especially social ethicists, are a primary audience but this book will also be of interest to social scientists focusing on religion, specialists in business ethics and human rights law, and scholars in development, environmental, and peace studies. Scholars can find in this book grist for the research mill, grist harvested from insights and reflection on the lived experience of Catholic peacebuilders who have been laboring in the field. Practitioners should also find this book helpful. It offers lessons from Catholics addressing mining and conflict around the world. It can also provide a theoretical edifice to support and strengthen ethical analysis, advocacy, and other forms of engagement on mining, and can indicate ways in which ecology, human rights, development, and peace can be woven together and better incorporated into the thinking and practice of the worldwide Catholic community.

R. Scott Appleby (2010, 19) has observed that "It is not enough ... for Catholics to expect secular experts to understand the theology of Catholic peacebuilding. Catholics must provide translation services." We recognize that Catholics engaging in peacebuilding, around mining and otherwise, are always doing so in concert with a range of religious and secular actors from varied backgrounds and value commitments. Those coalitions are essential for effective engagement on mining. This book offers some of that needed bridging by which other religious actors, scholars, policymakers, corporate executives, and social activists can better understand and perhaps appreciate the value of a Catholic peacebuilding approach to mining. That includes a deep well of ethical reflection and a vocabulary and framework for analyzing and critiquing injustice, malfeasance, and violence; dedication to an integral approach that tries to hold in balance all the various dimensions of the problems associated with mining; spiritual and pastoral resources that can offer distinctive forms of support for reconciliation, conflict transformation, social cohesion, and solidarity; rootedness in and commitment to impacted local communities; and a transnational institutional network that can dialogue and advocate in ways commensurate with the global dimensions of mining. At the same time, it reinforces the need for the Catholic community to understand, appreciate, and collaborate with other religious and secular scholars and practitioners in developing approaches to mining that are more just, peaceful, and sustainable.

Notes

1 See http://cpn.nd.edu.
2 For example, Peppard and Vicini (2015) and Opongo (2018) are two works that address extractive industries and natural resources from a Catholic ethical perspective, but only minimally cover mining specifically.

3 International Catholic organizations working on mining include the Justice, Peace, and Integrity of Creation Commission of the USG-UISG (https://www.jpicroma.org/), CIDSE (https://www.cidse.org/), and the Mining Working Group (https://miningwg.com/). Regional and national examples include Iglesias y Minería in South America (http://iglesiasymineria.org/), REBAC in Africa (https://rebaccongobassin.org/), CEPAS in the DRC (https://www.cepas.online/), and Alyansa Tigil Mina in the Philippines (https://www.alyansatigilmina.net/).

4 William P. George (2019) develops innovative theological ethics for mining and examines many different cases. But conflict and peace are not central concerns of his study, nor does he consider the roles and capacities of diverse church actors. Rigobert Minani (2017) examines injustices around mining but does not incorporate Catholic social teaching.

5 Though it is not exactly a Non-Governmental Organization (NGO), evaluating the Catholic Church as a transnational institution contributes to a need for additional research on the role of NGOs in mining since the church often functions like one and works closely with many Catholic and secular NGOs. According to Deonandan and Tatham (2016), there is a need for a better critical understanding of the contributions that NGOs can make, and in the Catholic context, the principle of subsidiarity could be a valuable tool for mitigating problems of cooptation that they note can often arise as NGOs work with local communities on mining (280–81).

References

Appleby, R. Scott. 2010. "Peacebuilding and Catholicism." In *Peacebuilding: Catholic Theology, Ethics, and Praxis*, edited by Robert J. Schreiter, R. Scott Appleby, and Gerard F. Powers, 3–22. Maryknoll, NY: Orbis.

Carozza, Paolo G., and Clemens Sedmak, eds. 2020. *The Practice of Human Development and Dignity*. Notre Dame, IN: University of Notre Dame Press.

Cichos, Katarzyna, Jarosław Sobkowiak, Radosław Zenderowski, Ryszard F. Sadowski, Beata Zbarachewicz, and Stanisław Dzieoński, eds. 2021. *Sustainable Development Goals and the Catholic Church: Catholic Social Teaching and the UN's Agenda 2030*. London: Routledge. 10.4324/9781003053620.

Deonandan, Kalowatie, and Rebecca Tatham. 2016. "The Role of Women and Non-Governmental Organisations in the Resistance to the New Extraction in Latin America: The Unexplored Dimensions." In *Mining in Latin America: Critical Approaches to Extraction*, edited by Kalowatie Deonandan and Michael L. Dougherty, 273–283. London: Routledge. 10.4324/9781315686226.

DiLeo, Daniel, ed. 2017. *All Creation is Connected: Voices in Response to Pope Francis's Encyclical on Ecology*. Winona, MN: Anselm Academic.

Finn, Daniel K., ed. 2021. *Business Ethics and Catholic Social Thought*. Washington, DC: Georgetown University Press. 10.2307/j.ctv1k03fz4.

Francis. 2019. "Address to Participants at the Meeting Promoted by the Dicastery for Promoting Integral Human Development on the Mining Industry." May 3, 2019. http://www.vatican.va/content/francesco/en/speeches/2019/may/documents/papa-francesco_20190503_incontro-industria-mineraria.html.

George, William P. 2019. *Mining Morality: Prospecting for Ethics in a Wounded World*. Lanham, MD: Lexington Books.

Hawksley, Theodora. 2020. *Peacebuilding and Catholic Social Teaching*. Notre Dame, IN: University of Notre Dame Press.

Ilo, Stan Chu. 2014. *The Church and Development in Africa: Aid and Development from the Perspective of Catholic Social Ethics*. Eugene, OR: Wipf and Stock.

Minani, Rigobert, ed. 2017. *Unmet Promises of Extractive Industries in Africa: Corporate Social Responsibility: Case Study of Kitwe (Zambia) and Katanga (DRC)*. Nairobi: Paulines Africa.

Opongo, Elias, ed. 2018. *Extractive Industries in Africa: Addressing Conflicts and Integrating Sustainable Development*. Nairobi: Paulines Africa.

Peppard, Christiana Z., and Andrea Vicini, eds. 2015. *Just Sustainability: Technology, Ecology, and Resource Extraction*. Maryknoll, NY: Orbis.

Rebman, Robert. 2020. "Corporate Responsibility, Catholic Social Teaching, and the Common Good: Reporting, Accountability, and Stakeholder Action." *Journal of Catholic Social Thought* 17, no. 2: 99–117. 10.5840/jcathsoc20201718.

Scheid, Daniel P. 2016. *The Cosmic Common Good: Religious Grounds for Ecological Ethics*. New York: Oxford University Press. 10.1093/acprof:oso/9780199359431.001.0001.

Schreiter, Robert J. 2010. "A Practical Theology of Healing, Forgiveness, and Reconciliation." In *Peacebuilding: Catholic Theology, Ethics, and Praxis*, edited by Robert J. Schreiter, R. Scott Appleby, and Gerard F. Powers, 366–397. Maryknoll, NY: Orbis.

1 Mining and peace: a scriptural reflection

Cardinal Peter K. A. Turkson

This book, born out of a larger project on mining and integral peace, addresses a fundamental question that has marked humanity's life and growth for decades, centuries, and millennia. Indeed, in terms of the human being's relationship with the earth, besides agriculture which provides nourishment for human existence, mining and mines dominate human activity on the land, providing material for tools, instruments, building, and decoration. So indicative were metals of human ingenuity and technological growth in the past that they gave their names to the different ages and periods of civilizations in human history. Thus, there was a movement from a Stone Age through a Copper Age, a Bronze Age, and an Iron Age. Classical literature also referred to a Silver Age and a Golden Age. Each of the different ages of history was characterized by the prevalent object or metal in use, even if it must be recognized that this development and history of the discovery of metals did not happen homogeneously in all geographical areas. Nevertheless, and with the necessary caveat, it is not wrong to assert that, from its beginning, human life has been associated to varying degrees with the discovery, production, and use of minerals and metals. And since not all places on earth are equally endowed with these deposits, mining gave rise to vigorous trade early in human history. Trade in minerals and metals continues to our own day, since we cannot do without them, even if the industry is fraught with tensions and conflicts between communities and businesses over land-use, and is marked by an ambiguity of purpose in the dignity-deficit with which it tends to leave communities and the environment.

In what follows, and by way of a prelude to the essays in this book, I shall modestly and briefly present the way this industry and the minerals which give rise to it are used to present phases of religious history in the Bible and some of its fundamental teachings, such as the transcendental character of God's covenant law, as *wisdom,* and the perfection of his presence, as pure gold, through redemption in Christ.

DOI: 10.4324/9781003094272-1

The wealth of creation and the goodness of the "land of promise"

In the second account of creation (Gen. 2:4bff.), God creates a garden, which was destined to be the home of Adam, whom God created from the dust of the earth and made a *living soul.* The river that waters the garden breaks up into four rivers on leaving the garden. One of these tributaries, called Pishon, is reported to be flowing through a land endowed with minerals: "The name of the first is the Pishon; it winds through the entire land of Havilah, where there is gold. The gold of that land is good; bdellium and onyx stones are there" (Gen. 2:11–12).[1] Since the mention of the other three rivers and where they flow are not followed by any indication of their properties, some commentators consider the mention of gold after Havilah to be a later addition. The canonical form of the account, however, would have us recognize that the earth (God's creation) did not only consist of vegetation, animals, birds, fishes, and the human person. God's creation (the earth) also contained precious minerals: a notice that prepares for the metallurgical work of Tubal-Cain (Gen. 4:22). Thus, if the second account of creation may be related with the first account (Gen. 1–2:4a) and considered "*good*," then the creation of an earth endowed with minerals and different metals, including those, like uranium, which can be used in nuclear weapons but also for peaceful purposes, is part of God's good purpose for the life of humanity on earth.

But in the general scheme of things in the Bible, the story of the creation of heaven and earth in the Book of Genesis, as the setting for human existence and of humanity's relationship with God, leads to the creation of a *new* heaven and a new earth in the Book of Revelation, as the goal and the fulfillment of history through redemption in Jesus Christ. Thus creation, with all its endowments and in all its goodness, quickly came to languish and to live under a curse for the sins of its inhabitants (Is. 24:3ff.). Since then, creation "awaits in eager expectation for the revealing of the children of God; for the creation was subjected to futility...in hope that creation itself will be set free from its bondage to decay and will obtain the freedom of the glory of the children of God" (Rom. 8:19–21). The earth and its endowments now exist under the stewardship of humanity in bondage to sin; and, as Pope Francis (2015, §2) observes, it is "the violence present in our hearts, wounded by sin," that is "reflected in the symptoms of sickness evident in the soil, in the water, in the air and in all forms of life." Living in bondage to sin, the human person's relationship with the earth and its resources, and his/her treatment of them, reflect his/her bondage, which is essentially the lack of justice towards God, towards one another, and towards creation. Set within this relationship, mining can reflect this bondage to sin; but it can also reflect the industry's response to a call to share in the vocation of humanity to redemption and to freedom to live in holiness and justice. (cf. Lk. 1:75).

Subsequently, the stories of the patriarchs attest to the possession of precious minerals as a sign of wealth (Gen. 13:2) and their use for commerce and transactions (Abraham's purchase of land [Gen. 23:16]; Jacob's purchase of land [Gen. 33:19]). The Exodus story continues to attest to the use of precious minerals as a symbol of wealth. The Israelites receive gold and silver from their neighbors on the eve of their departure from Egypt (Ex. 11:2), which will contribute to Aaron's making of the golden calf (Ex. 32) and the decoration of the Tabernacle (Ex. 35:24ff.).

But reference to the earth's endowment with mineral wealth comes up again in God's presentation of the "land of promise" to Moses: a land with wheat and barley, vines and fig trees, pomegranates, olive oil, and honey; a land where bread will not be scarce and you will lack nothing; a land where the rocks are iron and you can dig copper out of the hills (Dt. 8:9, cf. too, 8:13). The minerals, like the fertility of the land, represent the *goodness* of the promised land; and they are a covenant gift of God and a sign of God's gratuitous love, which is meant to be received in grateful obedience and fidelity. The warning about prostituting God's covenant gift is severe! (Dt. 8:13–14).

Postponing for a later consideration the religious values of these minerals, we may note here that the warning about abusing the good things of the land of promise is linked to an invitation to reckon also with the purpose of the use of created goods. Created goods are bestowed on the human family, especially in the context of the covenant, as an expression of God's benevolent goodness, and they are meant to be received with gratitude and in faithfulness to God, and used with concern and care for the rest of creation.

Mining and the contrast between earthly treasure and divine wisdom (Job 28)

The notices above about the endowment of creation and the land of promise with precious minerals and metals are complemented in Israel's pre-exilic and post-exilic histories by a lot of data about the abundant use of precious minerals in financial transactions, in temple and palace decorations, in votive offerings, and in household utensils.[2] The golden age of Israel's history, the reigns of David and Solomon, came during the Iron Age in world history. It was a period of great opulence and accumulation of wealth. David stored gold and precious stones for Solomon's building of the temple of the Lord (1 Chr. 22). The Queen of Sheba testified to the opulence of the palace of Solomon and the temple, and bestowed on him more wealth (1 Kg. 10; 2 Chr. 9). Biblical archeologists identify Eilat, the southern tip of the Dead Sea in the Jordan valley, as an attested mining site in ancient Israel—but it was for copper. Accordingly, the gold, silver, and precious minerals in the palaces and in the temple in ancient Israel (2 Kg.23:33, 35; 24:13–14; 25:13–16; cf. Dn.5:1–4) must have arrived there from tributes, gifts, and trade.

But it is Job 28 that provides a rare disclosure of the art and skills of mining in ancient Israel and in the Middle East. This passage affirms that Wisdom is divine, and is inaccessible to humanity despite its great efforts and marvelous skills, such as those it displays in mining precious minerals. Human success at mining precious minerals, as hidden and remote as they are, does not suffice to give the human person access to Wisdom, because Wisdom transcends human efforts and labor. To make this point, Job 28 refers to the mining of silver, gold, iron, and copper: "Surely there is a mine for silver, and a place for gold that they refine. Iron is taken out of the earth, and copper is smelted from the ore" (Job 28:1–2); and proceeds to describe the details of shaft mining: "Man puts an end to darkness and searches out to the farthest limit the ore in gloom and deep darkness. They open shafts in a valley ... they sway suspended ... They put their hand to the flinty rock, overturn mountains by the root" (Job 28:3–9). The challenges of this activity to the environment, to development, and to the dignity of the miners are as real in this passage as they are today.

With reference to mining and development and the dignity of people, it is very striking how the conduct described in Job 28 readily calls to mind the words of Pope St. Paul VI (1967, §25) in his encyclical letter on development, *Populorum Progressio*: "By dint of intelligent thought and hard work, man gradually uncovers the hidden laws of nature and learns to make better use of natural resources. As he takes control over his way of life, he is stimulated to undertake new investigations and fresh discoveries, to take prudent risks and launch new ventures, to act responsibly and give of himself unselfishly." What Paul VI says here about industrialization and technological growth applies also to the mining industry; and the challenges which he goes on to identify, as bedeviling industrialization with a human face, also bedevil the mining industry, causing tensions and conflicts with local communities. The challenges which the pontiff draws attention to are "profit as the chief spur to economic progress, free competition as the guiding norm of economics, and private ownership of the means of production as an absolute right, having no limits nor concomitant social obligations" (§26).

But Job's message is clear: despite the preciousness and great value of the minerals and the skills displayed in their mining and acquisition, they remain *created* gifts, *created* activities, and *created* skills, incomparable with and not to be substituted with Wisdom (Job 28:12. cfr. Prov. 8:10–11) and the search for it. As a divine attribute, Wisdom and the path to it (the fear of the Lord) are inaccessible to humans (Job 28:12ff.; cf. Baruch 3:15ff.). It is bestowed only as *gift!* Solomon prayed for it (1 Kg. 3:9), and God bestows it gratuitously on his people (Baruch 3:36–4:4).

The gift of gold, frankincense, and myrrh to baby Jesus

The three men whom the Gospel of Matthew calls "magi" (Mt. 2:1–12), came from the East on seeing the newborn king of the Jews' star in the east or at its rising (Mt. 2:2). Helped on by the word of prophecy to get to Bethlehem, they offer the baby Jesus gifts which are consistently described as gold, frankincense, and myrrh. While the three gifts were valuable items traditionally offered to a deity in ancient times, their presentation to Jesus symbolizes the magis' beliefs about the Christ-child's identity and his mission. Frankincense, used in worship, is offered to Jesus in recognition of his divinity. Myrrh makes us think of the "myrrh and aloe" which Joseph of Arimathea and Nicodemus brought for Jesus' burial (Jn. 19:38–39), and it is in recognition of Jesus' death, as messianic and redemptive. Gold, which is rust-free and non-corrosive, is offered in recognition of the kingship of Jesus, which, according to prophecy (Mt. 2:6; Lk. 1:33; Is. 9:7), is lasting, enduring, and eternal.

Metals and stones in the new creation: the splendor of the new heaven and new earth (Rev. 21)

The account of creation in Genesis, with which human history on earth begins, concludes with the preparation of a garden in Eden to be the home of Adam (joined later by Eve) whom God has created. This garden home was endowed with mineral wealth (Gen. 2:11–12), as we have seen. The end of human history is presented in the Book of Revelation as a new beginning of human history in a new heaven and a new earth, since the first heaven and the first earth have passed away (Rev. 21). Just as at the first creation (Gen. 1–3) God prepared a garden home for Adam, so does God prepare a new home, the holy city, the new Jerusalem (Rev. 21:2) for humanity redeemed: "those whose names are written in the Lamb's book of life" (Rev. 21:27).

What attracts our attention at this point is the presence or the place of precious stones and metal in the presentation of the two dwellings: the garden home of Adam and the holy city of humanity redeemed. While in the first (Gen. 2:10–13), the river Pishon flows around the whole land of Havilah, where there is gold, bdellium, and onyx, in the holy city of God's redeemed, the city itself and its street are gold, and its wall is built of jasper, whose foundations are "adorned with every jewel..." (Rev. 21:18–21). Adam's garden home is endowed with and possesses gold and precious stones; the holy city that comes down from heaven, by contrast, is wrought in gold and is gold itself. Thus, while in Eden gold has the sense of a mineral resource deposit and an endowment of a terrain that can be mined, in the holy city from heaven, which is gold and which is adorned with "the glory of God and a radiance like a very rare jewel, like jasper, clear as crystal" (Rev. 21:11), gold cannot have the sense of a mineral endowment to be mined. The holy city, which is gold and whose wall is built of jasper..." (Rev. 21:18–21), is also "a bride adorned for her husband" (Rev. 21:2). The holy city is the "bride, the wife of the Lamb" (Rev. 21:9). Thus, the gold of the holy city cannot have the physical sense of mineral deposit; it must have the sense of a symbolism, whose meaning we may now briefly explore.

The golden adornment of the tabernacle and the temple prefigures the holy city

At the behest of God (Ex. 25), Moses took donations of gold, silver, bronze, textiles, skins, and acacia wood from the people to build the ark of the covenant, the table of the bread of presence, and the lampstands in the tabernacle; and, as the dwelling place of God among his people (Ex. 25:8), the content of the tabernacle was wrought in pure gold. When wood was used, it was overlain with gold.

Similarly, when Solomon built the temple of the Lord, in the fourth year of his reign, as God's dwelling (1 Kg. 6:13), he overlaid "the whole house... the Cherubim, the floor of the house with gold (1 Kg. 6:22, 28, 30). For, as the metal that neither rusts nor corrodes, as already observed, gold was not just a store of value; it signified *lasting* value. Accordingly, when kings did accumulate gold, it gave expression to their splendor, their power and might. But, most importantly, they also sought to appropriate for their reigns the metal's attribute of a *lasting* permanence. When, however, the precious metal is predicated of God and of his presence, the unchanging and lasting character of the metal, as well as its brilliance, finds its worthiest symbolism, expressing the eternity, enduring perfection, and the glory of God.

Thus, the "holy city: the bride of the Lamb" of the Book of Revelation is pure gold, because it has been made perfect and eternal by Christ's passion.[3] It is not of any man's making. It is pure gold by reason of its sharing in the nature of the divinity that dwells there. Secondly, and by contrast with its prefiguration in the Old Testament, as well as other instances of human accumulation of gold as wealth, we need to distinguish between *being* something and *having* something. On earth, gold is possessed (having); and it is used as such. In the new creation, the holy city is gold (being), namely, completely invaded and filled with divine presence. The latter is an expression of the nature and essence of the holy city, where the purity of the minerals expresses the fullness of God's presence. The former instance of "having" entails an invitation, indeed, a vocation, to pass from "having" to becoming: to "being" the value and quality of what is possessed.

Conclusion

The symbolic use of precious metal in the Scriptures ascribe to the metal a character and a value that transcend its earthly value and usage, namely, as mere source and instrument of wealth. The Scriptural usage of the metal and others in its family appeals to a character of the metals which underlie their earthly usage, as valuable, precious, and lasting. But, ironically, these characteristics of the precious metals make them outlive the human person, who cherishes them and amasses them without, however, being able to share in their non-corrosive and incorruptible nature. Gold and the precious metals attain their true destiny, as it were, in the Book of Revelation, when they express the truly eternal, non-corrosive, and incorruptible kingdom of God.

In this sense, can the industry that seeks with such skills and dexterity to possess these precious metals be guided by some other objective and vision, besides that of possession or having wealth on earth? Can the industry also be inspired and guided by the values of the kingdom of God, where these precious metals are truly at home, as it were, giving expression to the heavenly Jerusalem, the holy city of God's redeemed, as the Scriptures suggest? Can the industry adopt values of the kingdom of God as guiding principles: the "common good," the "universal destination of the goods of the earth," "peace and justice," "concern for the poor," and "care of creation, our common home"?

Notes

1 Biblical references are from *The New Oxford Annotated Bible*, edited by Michael Coogan (2001).

2 Cf. Gen 23:16; 33:19; Ex 38:25; Lev 27:25; Jdg 8:26; 1 Sam 9:8; Ezek 45:12; 1 Chr 21:25; 29:7; Ezra 2:69; Neh 5:15; 7:70; Tob 5:15; Job 42:11; Mt 10:9.29; 17:24; 18:24; 20:13; 21:12; 22:19; 26:15; Lk 10:35; 12:6; 19:13; Acts 19:19

3 The idea of proving the worth of one's faith and fidelity to God through testing and trials is well attested in Scriptures (2 Chr 32:31; Jer 9:7; James 1:12; 1 Pt 1:6–7; 1 Pt 4:12–13; 1 Pt 5:10; Rev. 3.10). Often, the purification, testing, and the assaying of precious metals, like gold and silver, provide analogies for the testing of faith. Faith, tested and proven, is analogically presented as gold or silver tested and purified by fire (Ps 66:10–12; Prov 17:3; Job 23:10; Zech 13:9; 1 Pt 1:6–7; Rev 3:18).

References

Coogan, Michael, ed. 2001. *The New Oxford Annotated Bible*, New Revised Standard Version with the Apocrypha. New York: Oxford University Press.

Francis. 2015. *Laudato Si'*. http://www.vatican.va/content/francesco/en/encyclicals/documents/papa-francesco_20150524_enciclica-laudato-si.html.

Paul VI. 1967. *Populorum Progressio*. http://www.vatican.va/content/paul-vi/en/encyclicals/documents/hf_p-vi_enc_26031967_populorum.html.

Section I

Catholic engagement on mining in conflict zones

2 Religion and extractive industries: ethics, practice, and engagement

Katherine Marshall[1]

Religious engagement: framing patterns and legitimacy

Extractive industries present a classic challenge: they can be a mainstay and a powerful engine for good in an economy and society, but they can also be a source of distortions, tension, and conflict, to a point that the term "curse" is often applied to an abundance of such resources. Policies towards extractive industries are thus keenly debated in development strategies. A particular concern is that extractives have significant bearing on many conflicts; they are often linked to political, intra-state, and inter-state conflicts that surround the control and distribution of resources, land, and the revenues these resources produce. In some instances, directly or indirectly, religious tensions and conflicts are also involved.

The complexity of the diverse roles that extractive enterprises play is exemplified by how mining can be plotted to each of the UN Sustainable Development Goals (Columbia Center on Sustainable Development et al. 2016, 5). The many controversies around extractive industries include environmental challenges, physical scars to landscapes, inequitable distribution of benefits, propensities to large-scale corruption, and the impact on people directly and indirectly affected, including abuses of their human rights. As mining ventures extend deep into areas where indigenous communities live, their voices, rights, and ways of life are of central concern. Deep-sea mining also involves issues around disruption of the environment, and law and international justice. Each of these topics affect and involve religious communities.

Extractive industries rarely involve religious institutions directly. However, distinctive aspects of both extractive policy and operations have engaged religious actors in varying advocacy and negotiation efforts. Effects of mining on communities and relationships between communities and mining companies in many regions have spurred religious activism ranging from social and spiritual support to participation in active protests against, notably, large mining operations and government support for them. Ancient traditions of artisanal mining can also give rise to tensions and conflict. The significance for national and community welfare

DOI: 10.4324/9781003094272-2

of extractive policies and operations has led to active dialogue, analysis, and advocacy related to pertinent policy issues, such as taxation and environmental standards. Individually and collectively, religious actors can be powerful witnesses about the impact of mining, sharing knowledge and experience and highlighting vital ethical aspects. Religious peacebuilding, including direct mediation, is particularly significant in fragile and conflict-prone states, where many of the most controversial extractive investments are concentrated. Their actual and potential roles draw on their presence in mining areas, the trust they enjoy, and their willingness to highlight abuses of human rights and the vulnerability of affected communities.

This chapter explores evolving debates on extractive industry approaches and relevant policies, global and national, from the perspective of a range of religious actors who may speak from a specific tradition or community, or in an interfaith or intra-faith context. It situates various forms of involvement of religious actors, including the Catholic Church and interreligious and intra-faith bodies, within the broader, evolving policy debates about the roles of extractive industries in development strategies. Its focus is primarily on global forums, referring to country experiences to illustrate religious actor engagement. Driving questions include: How are concerns translated into action? Where has the impact of religious advocacy and action been most effective? And what are likely and productive avenues for the future?

Systematic frameworks for religious engagement in these debates are not clearly defined. Indeed, as with many development issues, efforts by religious actors to be part of global and national policy discussions can be met with skepticism as to their legitimacy in these settings. Demonstrating relevance and legitimacy as a stakeholder is thus a common challenge. Recent explicit focus by Catholic Church entities and the Church of England are prominent examples of both interest and capacity to engage. Debates and controversies extend from very local levels to transnational and global ones. Each situation is distinct, and religious engagement at local and national levels shows wide variation. It may address new mining ventures, distribution of resources and revenues, or community impacts such as forced resettlement and contaminated water. Efforts to address extractive industry policies at global and transnational levels also vary, both as to institutional engagement and positions on specific issues. Cases of systematic and effective religious engagement on national policies are quite rare, however, and overall suggests rather fragmented approaches. There is ample room for more cooperation and informed engagement in policy debates.

Evolving policy landscapes

Debates about extractive industries and their roles in development have evolved significantly in recent decades. Various broad trends in global agendas have tended to sharpen the focus on both opportunities and conflicts where mining ventures are concerned. The debates have been shaped by broader shifts in approaches both to development strategies and to the roles that extractive industries play in country and transnational strategies. Conflicts, especially with multinational mining companies, have contributed to shifts in approach and prompted engagement and reflection. These shifts have in turn affected the evolving engagements which have taken many forms between religious and development actors. These topics are summarized briefly here.

Corporate social responsibility and business practices

The spotlight for international development has turned increasingly to the ethics and practices of private business, reflecting in part the rapidly expanding roles of private finance for poor country development. Elaboration of approaches to Corporate Social Responsibility (CSR) has been one response, ranging from local charitable works to a full alignment of a business to social goals. Raymond Offenheiser's chapter takes up the latter in its argument that church actors should focus on advocacy to make mining companies into genuine development partners that engage in a robust and substantive way with communities' development goals. Business and Human Rights (BHR) represents a parallel development, of which Douglass Cassel's chapter is a good example. BHR reflects both internal and external pressures on companies to respond to concerns exemplified by extractive industry debates and tensions. A continuing issue is the voluntary nature of CSR and BHR (see Ramasastry 2015; Carter 2017; Gamu and Dauvergne 2018). A significant framework is the UN Global Compact, which was launched in 1999 and defines principles addressing corporate responsibility in a broader context. Companies are to measure conduct against international law, though without binding or enforceable rules. In practice, while the frameworks are far better articulated than in the past, with admirable principles as the foundation, governance, and enforcement of laws remain the responsibility of the relevant state.

Corruption and governance

The large resources involved in mining enterprises and often obscure financial flows have linked extractive industries to broader issues around transparency, corrupt practices, and capital flight. Concerns include the realities of substantial corruption associated with mining ventures and awareness that imbalances of power and various interests at work mean that host governments and especially communities get a raw deal, benefitting little from lucrative agreements. Global integrity movements aimed at bringing facts to light include the Extractive Industries Transparency

Initiative (EITI) (see Milin 2016),[2] Publish What you Pay (see van Oranje and Parham 2009),[3] as well as broader initiatives by global integrity alliances, notably Transparency International.[4]

EITI is of particular note, reflecting transnational efforts to make information more transparent, and set and enforce standards for good governance of oil, gas, and mineral resources. Established in 2002, it requires disclosure of information along the extractive industry value chain: how licenses and contracts are allocated and registered; who the beneficial owners of those operations are; the fiscal and legal arrangements; how much is produced; how much is paid; where the revenue is allocated; and contributions to the economy, including employment. EITI now includes fifty-five countries, each expected to publish an annual report disclosing information on contracts and licenses, production, revenue collection, revenue allocation, and social and economic spending. A quality-assurance mechanism is scheduled for at least every three years to assess performance towards meeting the EITI Standard and promote dialogue and learning at the country level. Efforts to hold all EITI-implementing countries to the same global standard is the key mechanism, and each implementing country has a national secretariat and multi-stakeholder group made up of representatives from the country's government, extractive companies, and civil society. An international multi-stakeholder board oversees the standard from a base in Oslo, Norway. The weakest element is that EITI depends on voluntary disclosure. Challenges include establishing consistency across national and international standards and meeting compliance costs.

Among issues that the EITI community has taken up is gender. This involves explicit assessment of the impact on women and girls of mining ventures and, more broadly, the benefits of diversity (or negative impact of a lack thereof). Women and girls bear a disproportionate share of the negative social, economic, and environmental impacts of extractives, and the goal of enhancing women's participation in decision-making around the management, development, and use of natural resources is now included in the recently revised EITI Standard (Thévoz 2019).

Principles of engagement with local communities and civil society organizations, including safeguarding policies applicable to multilateral financing organizations

A marked shift in development approaches concerns the roles of non-government actors (civil society) and approaches to engagement with communities. From an early tendency to rely on expertise, often external, with no formal mechanisms for engaging communities, practice shifted first to consultation, then to participation, and, today, to what is often termed empowerment, reflecting a conviction that local communities have both rights and wisdom in determining actions for their development. Such practices are expected and often required, for example in development strategies elaborated by development partners and in the poverty-reduction strategies that form part of debt-relief packages. The multilateral investment banks have been a focus of much discussion because, although their financing roles often represent a limited share of overall investments, their policies have tended to shape much of the dialogue around extractive policy approaches. For instance, the chapter by Holden and Montevecchio explains how World Bank mining policy in Africa influenced the Philippine Mining Act of 1995.

These processes have taken on special significance for communities that may not be fully integrated in national political systems, especially indigenous communities, where many tensions around extractive industries arise. This lies behind the effort to elaborate definitions and processes of "meaningful participation," notably through the development of standards and practices for free, prior, and informed consent (FPIC) (see Goodland 2004; Callies, Curtin, and Tappendorf 2003). These principles are enshrined in various international and national covenants, although there are continuing debates as to what each word represents and whether efforts to assure FPIC are indeed applied in practice.

Indigenous peoples

The rights of indigenous communities have particular significance for extractive industries debates, both because of egregious cases of abuse and because mining sites are increasingly being developed in areas occupied by indigenous communities. Thus, various efforts have been made to protect and assure the rights of these communities, shown in the chapters from Karl Gaspar on the Philippines and José Bayardo Chata Pacoricona on Peru. One milestone is the 2007 United Nations Declaration on the Rights of Indigenous People, which supports the principle that states should seek FPIC from indigenous groups regarding any measures affecting them.

Environmental protection

Concerns about environmental damage and the impact of fossil fuels on climate change have mounted over the decades and can now be seen as a central strategic concern for development strategies and approaches. This concern has highlighted links between mining practices and environmental protection. The environmental focus has sharpened over the years, with attention to safeguarding policies that assure full assessment and review of environmental impact of mining ventures, as well as protection of rainforests and biodiversity. These concerns link also to the land rights of affected populations as well as, for example, the negative effects of mining ventures on water supplies.

Fragility, conflict prevention, and peacebuilding

Many tensions and concerns about extractive industry practices and investments are centered in countries that are considered fragile, including those with violent conflicts. This reflects the association of poor governance and fragility, situations where abuses of power and neglect of sound policies and practices are accentuated. The concerns are heightened by the fact that tensions around natural resources can be a cause or accelerator of conflicts and can affect cost structures and risk assessments. Such tensions and competition involving extractive resources have often proved destabilizing. Over the past sixty years, an estimated forty percent of civil wars can be associated with natural resources; since 1990, at least eighteen violent conflicts have been fueled by the exploitation of natural resources. Rebel groups can exploit natural resources to fund war and competition can sharpen polarization. War displaces populations and refugee movements can degrade resource bases. Institutions designed to manage the environment are likely to be disrupted or shut down during a war or violent conflict (UN Interagency Framework Team 2012). This points to the importance of explicit attention to policies and practices towards extractive industries in the broader context of policies on fragility and its links to peacebuilding.

Religious engagement

Engagement on extractives between development institutions and religious actors, whether religious institutions, non-governmental organizations, or local communities, has been at best unsystematic and partial. In recent years this picture has changed, with a range of institutions and networks whose purpose is to build bridges. However, in many instances, religious actors do not have the assurance of a "seat at the policy table," whether in discussing a national law on mining or on decisions on financing or negotiation of contracts and agreements (Marshall 2021).

There is thus no accepted "place" for religious engagement in many of the international debates that affect extractive industry policies or in efforts to address problems that arise. The role that the Catholic Church plays, at both local and global levels, has grown because of local church experiences with mining in different countries, the active diplomacy of Vatican leaders, and recognition by transnational mining companies, the Majors in particular, of the moral issues involved and the trust that many communities have in the church and its leaders. Likewise, the extensive experience in the sector by the current Archbishop of Canterbury has contributed to his influence and credibility. Religious insight, experience, and engagement are pertinent for each of the topics highlighted above. A question and challenge for the future is how a more systematic and interreligious involvement might emerge and how it might be facilitated and encouraged. Pressures to go beyond rhetoric to operational programs and engagement in national development strategies have increased, and tensions reflected in polarization of politics and shrinking civil society space suggest that a well-conceived, collaborative, multireligious engagement might address some of the complex ethical, human rights, and practical challenges involved.

Extractive industry strategies and approaches

In recent years, various stocktaking efforts, official and academic, have sought to address the increasingly sharp debates about policies and practices for extractive industries with an eye toward their impact on sustainability and development. These continue, with, in a contemporary setting, a particular focus on climate change implications in the leadup to the November 2021 Glasgow COP 26 Summit. But an important early example of both review and tensions was the Extractive Industries Review (EIR) initiated by the World Bank in 2001–2003. It has been followed by numerous other stocktaking reviews, at transnational and national levels, but the essential tensions and difficult debates are illustrative of the polarization on the topic.

The EIR was established primarily in response to pressure from environmental non-governmental ogranizations (NGOs), particularly Friends of the Earth, on the World Bank to stop funding extractive projects. Its framing was shaped by the earlier World Commission on Dams (WCD), which focused on large dam projects which had attracted active protests. But unlike the WCD, the EIR focused on the World Bank Group's (WBG) engagement with the extractive sector. It was led by Emil Salim, ex-Indonesian Minister for the Environment and Population during the 1980s and early 1990s. Salim commissioned a series of fully independent indigenous case studies of the impact of World Bank extractive sector activities on indigenous peoples in Colombia, Cameroon, Papua New Guinea, India, Indonesia, the Philippines, and Russia. The report's review and finalization encountered a rocky path. A draft of the report, released in 2003, omitted inputs from indigenous and non-governmental organizations; however, it concluded that World Bank engagement with the sector was beneficial and should continue. Civil society and indigenous organizations threatened to pull out of the process altogether unless the process changed, ensuring that the final report would be inclusive, independent, and guided by a multi-stakeholder advisory panel that would include at least one indigenous expert. The final report included recommendations for reforms in World Bank policies on human rights and enforceable FPIC standards. The World Bank, however, did not accept many of the EIR's recommendations, notably on indigenous peoples and FPIC, advising that the Bank would only support extractive projects with free, prior, and informed *consultation* (not "consent") that resulted in the broad community support of affected communities.

Similar debates continue to this day around how well the World Bank and other multilateral banks have or have not implemented effective and informed participation of key stakeholders in projects and standards, including approaches to assuring protection of the rights of indigenous peoples (see World Bank 2021; Environmental Defense Fund 2004). Concerns expressed by indigenous peoples and civil society about problems with review processes and consultation procedures persist. Weak follow-up and commitment to act on the findings of sector reviews generate frustration and disillusionment among indigenous organizations and leaders who engaged with these processes. There is a clear need for global mining companies, economists, and development strategists to be more deliberately engaged with communities. Religious actors' witness on the impact of mining can thus have great value. Throughout these different phases and in different ways, religious actors have played growing roles and, in some places and from some perspectives, have had growing force.

Religious and interreligious approaches

The most prominent instances of religious engagement on extractive industry issues involve the Catholic Church. Since the Catholic Church is the primary focus of this volume, the following sections highlight initiatives of other religious communities and interreligious action, noting areas where religious involvement is weak or absent.

With the notable exception of the Interfaith Center for Corporate Responsibility (ICCR) and linked, more global efforts centered on the management of financial portfolios and engagement of shareholder power to influence policy and behavior, the wide variety of specific and primary actions by different religious groups are focused on specific projects and places. These include, for example, advocacy and protests by the International Network of Engaged Buddhists focused especially on mining activities in Myanmar and in the Philippines, courageous support by Arya Samaj leader Swami Agnivesh for Advasi communities protesting mining ventures in their areas of India, and interfaith efforts to address mining policies in Zimbabwe and Malawi.

The Church of England and the Anglican Communion more broadly have been especially active in policy and advocacy on extractive industries. The national investment bodies of the Church of England—the Church Commissioners, the CBF, and the Pensions Board—recently published the Extractive Industries Policy, in which they threatened to divest from extractive companies that fail to comply with its latest policy on ethical investment (Church of England 2017). The policy was the culmination of two years of theological reflection and research on the industry. Companies have a responsibility to uphold basic standards of human rights; avoid corruption and pay taxes; protect the environment and ecology; and monitor social and economic concerns, such as labor standards, community engagement, fair pay, and collective representation. The policy also bars investments in companies that extract on world heritage or protected sites, and warns against the dilution of human rights and health-and-safety standards where joint ventures are undertaken (Williams 2017).

A variety of ecumenical initiatives have focused on extractive industries. These include, within the broader context of its faith and justice work, a diverse set of events and studies by the World Council of Churches (WCC) and its affiliates. For example, there was a session focused on mining at the 2015 World Social Forum (WCC 2015). The president of the All Africa Conference of Churches opened an event with Alternative Mining Indaba in 2011 by highlighting the negative impact of mining both for the environment and for human dignity: "The extractive industries sector has acquired notoriety for being non-transparent and unaccountable. Most problems in the extractive industries are often in developing countries where minerals and oil are being extracted in communities and the proceeds do not benefit these communities" (Mokiwa 2011). And at a 2011 WCC gathering, moving statements were made on the evils of mining, especially in Tanzania:

> Open pit mining in Tanzania is destroying God's creation. Arsenic and heavy metals are contaminating water and soil. Health of people near mining areas is deteriorating. Skin diseases are increasing. Farmland is contaminated. Evictions from land have led to loss of human dignity. 400,000 small scale miners have been put out of work and many people have been displaced from their ancestral lands to make way for gold mines. Tanzania as a whole is not benefiting from gold mining. (Hughson 2011)

In Latin America, Iglesias y Minería, an ecumenical coalition of about seventy Latin American Christian organizations, was, from 2013, "united by the challenge of the impacts and violations of socio-environmental rights perpetrated by mining companies in the lands where we live and work." Believing in "the strength of popular organization in our lands, based on the intense work of our Christian leadership and on the spirituality and commitment of our faith communities" the group came together to organize due to "the growing criminalization and persecution of our leadership," whether on the part of the mining companies or national governments. A first meeting in Lima, Peru confirmed the importance of organization of the churches at the grassroots level, of exchange between Christian communities, and of the debate about these issues at stake (Iglesias y Minería, n.d.).

Approaches to the ethical and religious approaches to mining ventures in Islamic institutions and traditions merit a separate study, given extensive global and local experience and wide diversity among country situations. This experience is embedded in relationships between different states and religious bodies and evolving approaches towards, especially, care for the environment (see Zakaria 2020; Gade 2015, 2019). Of particular note is the rich experience in Indonesia, where a succession of fatwas on environmental matters, including mining, have grounded advocacy on religious law (Mangunjaya and Praharawati 2019).

A prominent interfaith effort, with a fifty-year history of activism, is the US-based Interfaith Center on Corporate Responsibility (ICCR). Initially established mostly by Protestant Christian leaders responding to apartheid in South Africa, it has more recently pioneered the use of shareholder advocacy to press companies on environmental, social, and governance issues. Catholic and Jewish groups have joined and the coalition today includes over 300 global institutional investors from different religious traditions, holding more than $4 trillion in managed assets. ICCR members include faith-based organizations, socially responsible asset management companies, unions, and foundations.

ICCR members engage company managements to identify and mitigate social and environmental risks resulting from corporate operations and policies. The fundamental proposition is that "responsible and sustainable business practices—and a strong corporate culture of ethics—are in the long-term interest of both companies and investors" (ICCR, n.d.). ICCR members conduct roughly 300 dialogues annually, with over 200 companies, working toward specific goals. ICCR also hosts industry roundtables that convene multiple companies and investors, and other relevant stakeholders like NGOs, community groups, and industry trade associations, to accelerate progress on specific issues.

ICCR has a mining caucus that focuses explicitly on extractive industry issues. The group focuses broadly on how the private sector, and, in some cases, in tandem or conversation with the public sector/government, can play a role in affecting the extractive industries and the forms of development that are taking place. A focus is the need (and the perceived current trend) to move towards more sustainable and ethical forms of development. By incorporating more voices, particularly those of indigenous people, the goal is to remain accountable and responsible to the impacts of the industry, past and present.

Since 2001, initially under the impulse of the UK-based Alliance of Religions for Conservation (ARC), models with some kinship to ICCR have focused on mobilizing the financial resources and shareholder power of faith institutions. The International Interfaith Investment Group (3IG) was established in 2005. While it accomplished some important foundational work, the institution failed. FaithInvest, launched in 2019, seeks, like its predecessors, to establish faith-consistent investment principles so that their investments support environmental and sustainable development, including in the mining sector.[5]

Religions for Peace, the global interreligious body, has engaged on extractive industries through local interreligious councils but has yet to adopt a global approach or focus. During the leadup to the August 2019 Global Assembly in Lindau, Germany, regional consultations were held on major focal areas that included Just and Sustainable Societies (Dennis 2020). However, neither in the regional nor the global consultations did extractive industry issues emerge as a central concern.

Aspects of extractive industry policies are relevant for the annual agendas of the G20, and thus the G20 Interfaith Forum has highlighted the work and roles of religious actors in shaping priority policy concerns, notably appropriate engagement with indigenous communities. A session at the 2019 G20 Interfaith Forum highlighted parameters of the debate, including the work of faith communities and leaders with local organizations and peoples to support resistance movements and challenge the mining industry in protecting the rights and livelihoods of the people most affected; philosophical/theological perspectives contesting greed and excessive reliance on technology; and an international investment and development perspective seeking to influence large companies and actors to shift approaches through increased corporate responsibility (Finn 2019). Addressing corruption in the industry has emerged as an area of focus (G20 Interfaith Forum 2020). Extractive industries were alluded to during the 2020 G20 Interfaith Forum and will be on the agenda for the 2021 Forum.

Concluding reflections

Natural resources have the potential to drive growth, development, and poverty reduction. The extractive industries sector plays a significant economic role in some sixty-three countries (International Monetary Fund 2012). However, many of these countries face challenges, such as resource dependency and weak governance, and many are fragile states. Various multilateral organizations work to support these countries in managing oil, gas, and mining in ways that contribute to sustainable growth and development, protect communities, and reduce carbon emissions. Major areas of focus are strengthening the transparency, governance, institutional capacity, and regulatory processes of countries' extractive sectors. These efforts have particular importance during the COVID-19 emergency, given the need for countries that rely on oil, gas, and mining to boost economic recovery and resilience and protect the poorest and most vulnerable, including those working in artisanal and small-scale mining.

The challenges involved loom large. Apart from longstanding and contemporary contestation around the optimal balance in extractive industry policies and specific challenges around human rights, many states and regions with underexploited mineral resources face intensifying conflicts accentuated by competitive pressures involving industry and local authorities. Conflict has expanded an illegal and unregulated extractive economy.

Common cause and action among religious communities on social justice issues, including extractive industries, is an ideal. This should form an integral part of peacebuilding strategies and action. However, practical efforts to translate such aspirations into practice are partial and, in many situations, limited, in part because of the diversity of local situations and different approaches to the complex issues involved. As is the case for the Catholic Church, most mobilization and advocacy is focused on specific cases in specific localities. Given the profound ethical issues involved and the strategic challenges that extractive industries present, building on the examples of constructive religious engagement and cooperation among different parts of religious communities and different traditions offers significant potential to contribute to more sustainable and equitable development strategies at national and regional levels.

Notes

1 I offer appreciation to Fr. Séamus Finn who provided valuable comments, and Gabrielle Mendelsohn for significant research support.
2 See https://eiti.org/.
3 See https://www.pwyp.org/.
4 See https://www.transparency.org/en.
5 See https://www.faithinvest.org/.

References

Callies, David L., Daniel R. Curtin, and Julie A. Tappendorf. 2003. *Bargaining for Development: A Handbook on Development Agreements, Annexation Agreements, Land Development Conditions, Vested Rights and the Provision of Public Facilities*. Washington, DC: Environmental Law Institute.

Carter, Russell A. 2017. "CSR Earns a Seat at the Boardroom Table." *Engineering and Mining Journal* 218, no. 12: 38–41, 72. https://www.e-mj.com/digital-editions.

Church of England. 2017. "Extractive Industries: The Policy of the National Investing Bodies of the Church of England and the Ethical Investment Advisory Group's Advisory and Theological Papers." https://www.churchofengland.org/sites/default/files/2018-03/Extractive%20Industries%20Policy%20and%20Advice.pdf.

Columbia Center on Sustainable Development, UN Sustainable Development Solutions Network, United Nations Development Programme, and World Economic Forum. 2016. "Mapping Mining to the Sustainable Development Goals: An Atlas." White Paper. World Economic Forum. https://resources.unsdsn.org/mapping-mining-to-the-sustainable-development-goals-an-atlas.

Dennis, Chelsea. 2020. "Organizing Religious Communities to Work for a Just Peace." *Nonprofit Quarterly*, March 9, 2020. https://nonprofitquarterly.org/organizing-religious-communities-to-work-for-a-just-peace.

Environmental Defense Fund. 2004. "Environmental Defense Calls World Bank Response to Extractive Industries Review All Talk and Little Action." August 3, 2004. https://www.edf.org/news/environmental-defense-calls-world-bank-response-extractive-industries-review-all-talk-and-littl.

Finn, Séamus. 2019. "Extractive Industries: The G20 Interfaith Forum Outlines Fundamental Debates and Dilemmas." *Berkley Forum, Berkley Center for Religion, Peace and World Affairs*, Georgetown University. June 30, 2019. https://berkleycenter.georgetown.edu/responses/extractive-industries-the-g20-interfaith-forum-outlines-fundamental-debates-and-dilemmas.

G20 Interfaith Forum. 2020. "Governance and Accountability: Fighting Corruption in the COVID-19 Emergencies." *Policy Brief*. October 13, 2020. https://www.g20interfaith.org/app/uploads/2020/09/2020_g20i_policy_brief_governance_and_corruption.pdf.

Gade, Anna M. 2015. "Islamic Law and the Environment in Indonesia: Fatwa and Daʿwa." *Worldviews: Global Religions, Culture, and Ecology* 19, no. 2: 161–183. 10.1163/15685357-01902006.

Gade, Anna M. 2019. *Muslim Environmentalisms: Religious and Social Foundations*. New York: Columbia University Press. 10.7312/gade19104.

Gamu, Jonathan Kishen, and Peter Dauvergne. 2018. "The Slow Violence of Corporate Social Responsibility: The Case of Mining in Peru." *Third World Quarterly* 39, no. 5: 959–975. 10.1080/01436597.2018.1432349.

Goodland, Robert. 2004. "Free, Prior and Informed Consent and the World Bank Group." *Sustainable Development Law and Policy* 4, no. 2: 66–74. https://digitalcommons.wcl.american.edu/sdlp/vol4/iss2/13.

Hughson, Greg. 2011. "Report on the World Council of Churches International Ecumenical Peace Convocation." https://www.otago.ac.nz/chaplain/otago021114.pdf.

Iglesias y Minería. n.d. "Quiénes Somos." Accessed May 24, 2021. https://iglesiasymineria.org/quienes-somos.

Interfaith Center on Corporate Responsibility. n.d. "About." Accessed May 24, 2021. https://www.iccr.org/about-iccr.

International Monetary Fund. 2012. "Macroeconomic Policy Frameworks for Reseource-Rich Developing Countries." August 24, 2012. https://www.imf.org/external/np/pp/eng/2012/082412.pdf.

Mangunjaya, Fachruddin Majeri, and Gugah Praharawati. 2019. "Fatwas on Boosting Environmental Conservation in Indonesia." *Religions* 10, no. 10: 570. 10.3390/rel10100570.

Marshall, Katherine. 2021. "Religion and International Development." In *Handbook on Religion and International Affairs*, edited by Jeffrey Haynes. London: Routledge (forthcoming).

Milin, Zorka. 2016. "Mapping Recent Developments in Transparency of Extractive Industries." *Business and Human Rights Journal* 1, no. 2: 321–326. 10.1017/bhj.2016.8.

Mokiwa, Valentine. 2011. Address at the Meeting of Alternative Mining Indaba, Cape Town. Quoted in Norwegian Church Aid Alliance, "African church leader challenges internatonal mining companies," February 3, 2011. Mines and Communities. http://www.minesandcommunities.org/article.php?a=10724

Ramasastry, Anita. 2015. "Corporate Social Responsibility Versus Business and Human Rights: Bridging the Gap Between Responsibility and Accountability." *Journal of Human Rights* 14, no. 2: 237–259. https://ssrn.com/abstract=2705675.

Thévoz, Indra. 2019. "Building a Movement for Gender Equality in Extractives." EITI.org. August 28, 2019. https://eiti.org/blog/building-movement-for-gender-equality-in-extractives.

UN General Assembly. 2007. *UN Declaration on the Rights of Indigenous Peoples*. A/RES/61/295. September 13, 2007. https://undocs.org/A/RES/61/295.

UN Global Compact. n.d. "Ten Principles of the UN Global Compact." Accessed May 24, 2021. https://www.unglobalcompact.org/what-is-gc/mission/principles.

UN Interagency Framework Team for Preventive Action. 2012. *Toolkit and Guidance for Preventing and Managing Land and Natural Resources Conflict. Extractive Industries and Conflict*. https://www.un.org/en/land-natural-resources-conflict/pdfs/GN_Extractive.pdf.

van Oranje, Mabel, and Henry Parham. 2009. *Publishing What We Learned: An Assessment of the Publish What You Pay Coalition.* London: Publish What You Pay. https://eiti.org/files/documents/Publishing%20What%20We%20Learned.pdf.

Williams, Hattie. 2017. "Church of England Targets Extractive Industries." *Church Times*, November 10, 2017. https://www.churchtimes.co.uk/articles/2017/10-november/news/uk/new-church-of-england-policy-targets-extractive-industries.

World Bank. 2021. "Extractive Industries: Overview." Updated April 26, 2021. https://www.worldbank.org/en/topic/extractiveindustries/overview.

World Council of Churches. 2015. "Churches Speak Against Negative Impact of Mining at World Social Forum." March 31, 2015. https://www.oikoumene.org/news/churches-speak-against-negative-impact-of-mining-at-world-social-forum.

Zakaria, Faizah. 2020. "From Fatwas to Ecumenism: Environmentalism in Southeast Asia." Berkley Forum, Berkley Center for Religion, Peace and World Affairs, Georgetown University. November 10, 2020. https://berkleycenter.georgetown.edu/responses/from-fatwas-to-ecumenism-environmentalism-in-southeast-asia.

3 The Church in Colombia and extractives: pastoral accompaniment using an eco-theology of peace[1]

Sandra Polanía-Reyes and Héctor Fabio Henao

This chapter describes and analyzes the interaction between conflict, extractives, and institutions at the intersection of civic participation and pastoral accompaniment in Colombia. For generations, mining has been part of life for many Colombians. After the 2016 accord, many, including the government, counted on mining to deliver a peace dividend. In an effort to accompany local communities, the Catholic Church has responded in a variety of ways to challenges posed by mining.

This study maps and analyzes that church response. The first section describes the current socio-political context of Colombia. After an internal social and armed conflict of more than six decades, the country is going through a phase of peace implementation and restoration, and faces challenges and coordination failures involving the current government, economic actors, and civil society. The second section addresses Colombia's legacy of mining and the trade-off between conservation and extractive-dependent development in a country with a large degree of biodiversity, a fraught history of violent conflict and social inequity, and large amounts of non-renewable resources.

The third section presents a national survey of all ecclesiastical jurisdictions examining the dynamics and impacts of mining, and the responses of civil society and the Catholic Church. It provides valuable data on the extent of mining that is not readily available from the government and other sources. A baseline was conducted in 2013 and a follow-up in 2020. The timeline allows us to analyze changes related to the national peace agreement of 2016. We find a major increase in mining throughout Colombia since 2013. Moreover, the social, environmental, economic, and cultural impacts of mining on communities have worsened. The fourth section analyzes the church's response to the challenges of mining and describes the resources and tools it has used when accompanying communities in reconciling mining, integral human development, and peace. We complement this section with reflections on how the church's response could be improved.

DOI: 10.4324/9781003094272-3

The socio-political context for pastoral accompaniment on mining

Colombia is framed by a great contrast between positive macroeconomic growth and an internal social and armed conflict of more than six decades. The presence of illegal armed actors, both of the extreme left and right wings, has generated a deep humanitarian crisis with more than nine million registered victims. In a country of 48 million inhabitants, this means that 18% of the total population of Colombia has been victims of the internal armed conflict, with the social, economic, and cultural consequences that this entails. The humanitarian crisis has also been worsened by the Venezuelan diaspora, which adds uncertainty and unrest (see Colombian Ministry of Foreign Affairs 2021).

The Catholic Church's pastoral accompaniment in Colombia takes place amidst one of the most complex and lengthy conflicts in the world. Two attempts in recent years to end the conflict have faced major difficulties. First, the negotiation process with the guerrillas of the National Liberation Army (ELN) that began in March of 2016 failed due to a terrorist act committed in Bogotá in January of 2019. Though the COVID-19 pandemic has renewed interest in a possible peace, formal peace talks have not resumed. Second, the 2016 Peace Agreement with the Revolutionary Armed Forces of Colombia (FARC) ended a five-decade conflict. However, there has been growing disapproval of the peace process by the current government and implementation of the agreement has been difficult. At the end of the fourth year of implementation, only 28% of stipulations have been fully implemented, 18% of stipulations are at an intermediate level of progress, and 54% of commitments have either begun but made marginal progress, or have yet to be initiated (Peace Accords Matrix Barometer Initiative 2021). The slow implementation has allowed illegal armed groups to reemerge in former FARC territories.

There are disputes over territory today between guerrillas, drug traffickers, paramilitaries, and other illegal armed actors who aim to defend their economic and political interests. These interests collide with the government's and are generally tied to drug trafficking and the legalization of mega-economic projects, such as mining, extraction of natural resources, agro-industry, extensive livestock farming, monocultures, and generation of electric power, in which the land and the abundant natural resources become strategic targets for national and transnational actors. These disputes increase human rights and international humanitarian law violations, as well as forced displacement, in most rural territories.

In particular, Colombia is the most dangerous country in Latin America for social leaders (UN Human Rights Council 2019, sec. 20). Human Rights Watch reports that between 2016 and February 2021 four hundred human rights defenders were killed (Cuartas 2021), although some sources report higher numbers.[2] Besides these assassinations, there have also been numerous cases of threats, displacement and confinement, robberies, kidnappings, attempted killings, sexual violence, disappearances, and confrontations. Most of the indigenous Afro-Colombian communities and campesinos who suffer the attacks are in areas where mining is prevalent.

The legacy of mining in Colombia

Mining in Colombia must be understood in the context of immense biodiversity. The country occupies 0.7% of the world's continental surface and has 10% of the world's biodiversity. But today, half of Colombia's ecosystems are critically threatened, and a third of its plants and half of its animals are under threat from habitat loss due to oil, mineral, and metal extraction, deforestation, extensive cattle-ranching, generation of electric power, wildlife trafficking, and the impact of the coca trade, among others.

Mining is the means of subsistence for a large number of Colombians. About six million people live in areas where mining occurs. The prominence of mining extends back far before the colonial era, and was a central aspect of the colonial program. Mining, including the abuses it brought and the problems that it left behind, is an irrevocable part of the nation's heritage, although the role it continues to play in the country's economic development is tremendously complicated. Violence and conflict, environmental degradation, and social injustice swirl around the industry.

Before the arrival of the Spaniards, indigenous people engaged in agriculture and in artisanal mining, especially of gold. These activities were part of their culture and tradition. After the conquest, the extraction of gold and silver intensified and became a way for indigenous peoples to participate in commercial exchange. By the nineteenth century, the country had the most mining activity in Latin America. But by the 1950s the low prices of precious minerals worldwide and the internal armed conflict caused mining activity to decrease to the point that by 1980 there were only two large national mining companies still in operation: Cerrejón and Cerro Matoso.

However, the oil and mining boom at the end of the 1980s and the beginning of the 1990s reactivated Colombia's extractive industries. In the last two decades, extractive industries have had so much support that the Uribe government (2002–2010) granted 8.53 million hectares, a territory the size of Austria, for exploitation. During the Santos government (2010–2018), mining became one of the five pillars of national development, which prompted a strong campaign to attract foreign investment through regulatory easing. However, this model is based on only the extraction of raw materials, without generating any added benefit to the communities or the environment. Other sectors have raised the alarm that economic growth of extractives without technological innovation or the generation of new industries will jeopardize Colombia's economy in the long term.

Today, there are three types of mining in the country: small-scale traditional or artisanal, medium-scale, and industrial or large-scale. Small-scale traditional mining has been used by indigenous cultures since ancient times and is also used by miners settled in small towns, some of whom are displaced. Although artisanal mining can frequently have serious negative consequences for the environment, a cooperative of ethnic communities in Choco called Oro Verde promotes use of a technique traditionally called "*mazamorreo*" or "*barequeo*" to extract gold, silver, platinum, or other minerals by hand, and the cooperative became a global pioneer in the movement for ecologically friendly artisanal mining. Oro Verde was credited as an inspiration for the global Alliance for Responsible Mining that promotes socially responsible and environmentally sustainable artisanal mining practices (Alliance for Responsible Mining 2014).

Medium mining uses some heavy equipment, like dredgers and backhoes, which allows greater productivity but has a greater environmental impact. That impact is made worse in some cases because of irregularities in the procedures and extraction processes and lack of compliance with national regulations. This type is important to illegal armed actors, especially in the areas of the Pacific and the Amazon, where they create alliances with national or foreign miners (Peruvians and Brazilians) to extract gold and wood in a disorderly and illegal manner. Several reports have documented that 80% of the mining in the country is illegal, and 60% of the illegal mining is in protected areas. This is due to the shortcomings of environmental licensing, the inefficient management of local governmental offices, and ineffective enforcement actions.

Large-scale mining, which is associated with megaprojects, is the type carried out by multinationals that benefit from favorable legislation and fiscal policies. These projects have a corporate social responsibility component, though that usually does not compensate for environmental damages or the serious negative social, cultural, and economic impacts on local communities.

Although current President Ivan Duque has called for "responsible development," the National Development Plan 2018–2022 introduced an expansion of the mining sector that has made the country one of the top extractive economies in Latin America. In order to achieve these goals, it was necessary to sell and/or concede significant territory to multinational companies, leading to an increased presence of large-scale mining operations. Some exploitation contracts with multinational companies involve protected areas. Mining projects are also planned in indigenous territories, such as a license to exploit 1,995 hectares by open-pit mining in the Great Indigenous Reservoir of Vaupés (Agenda Propia 2020).

History shows that there is no linear relation between conflict and mining, as conflict dynamics depend on the previous presence of the state, infrastructure, the type of mineral and extraction method, and the geostrategic importance of the jurisdiction for armed and criminal groups

(Rettberg et al. 2020). It is clear, however, that mining intersects and interacts with violent conflict in numerous regions. This includes medium mining operations and artisanal mining areas where armed actors vie for territorial control for illegal mining or coca cultivation, and large-scale mining where protest is criminalized and community leaders are victimized.

Surveying the church on extractives, governance, and peace

In the midst of this national mining panorama, the social struggles for the protection of the territory, the life of the communities, and the biodiversity of the country have started to become more public. Several ecclesiastical jurisdictions began to request support to analyze and accompany the communities affected by extractive activities. In 2013, the Episcopal Conference of Colombia and Caritas Colombia conducted a semi-structured baseline survey that examined the perceived effects of extractives in the jurisdictions. The study examined the existing relationships between the mining industry and the intensification of conflict, the exploitation and use of resources, the presence of illegal armed groups, and the displacement of communities. Seventy-four Catholic Church representatives participated. For the first time, a map of perceived effects of extractives in Colombia was available to the public.[3]

During the Colombian peace process (2012–2016), some hoped for dialogue about the role of extractives and peace, but the topic was avoided in order to achieve an agreement. Despite being one of the world's most comprehensive peace accords, the failure to address mining reflects a missed opportunity to connect integral peace, ecology, and development.

Given the need for reliable microdata showing how extractives were affecting the territories after the peace agreement, in 2020 Caritas Colombia developed a survey of representatives of the ecclesiastical jurisdictions in order to determine the main social, political, economic, cultural, and environmental impacts of extractives. This follow-up study improved on the 2013 survey in three ways. First, it allowed information to be collected and analyzed in real time, reducing the duration of data collection by 60%. Second, three modules were added: social capital, forms of pastoral accompaniment, and sources of doctrinal support from the Catholic Church. Third, the results of the 2013 survey were adapted to a database that allowed comparisons with other databases and with the 2020 data (Polanía-Reyes 2021). The 2020 survey was conducted with 104 church representatives (archbishops, bishops, and social pastoral directors) and included each ecclesiastical jurisdiction.

The survey confirmed that the extraction of natural, mineral, or hydrocarbon resources is an unstoppable reality. It occurs in 92% of the country's jurisdictions. This is an increase of 11% since 2013. There is also an increase in the perceived exploitation of iron ore, aggregates, nickel, limestone, coltan, and emeralds. When disaggregated at the sector level, most of the significant increase in extraction is for construction and industrial resources. Large-scale and artisanal exploitation constitute 68% of total extraction. In addition, the number of projects in an exploitation phase increased from 52% to 89%. There is also a 78% increase in extraction in protected areas compared to 2013.

As of 2020, there is no social engagement with the community in 49% of the extractives cases, and a high degree of engagement in only 8%. In many of the protected natural areas, areas of indigenous reserves, and/or collective territories of Afro-Colombians, companies do not follow the procedure of prior or popular consultation, violating a fundamental right of those communities. Despite the fact that the oil industry is a century old and oil generates the most income in the country, the industry has low levels of engagement with the communities and often breaches agreements. In total, companies have breached agreements in 58% of cases.

The survey found that the majority of mining communities have high rates of poverty and deterioration of the quality of life and the social fabric. Whether violence and conflict might explain this relationship is an open research question. Positive economic effects included revitalization of the local economy (50%), an increase in employment (49%), and enhancing local government revenues through taxes (30%). Negative effects included shocks to the local economy (62%), price increases (60%), and the limiting or prohibition of traditional mining (42%).

Perception of positive social effects deteriorated in 2020 compared to 2013. In 2020, 58% of reported projects did not present any positive social impact. The remaining projects present stronger presence of national institutions (20%) and defense manpower (15%), signaling less crime and violence. Many negative social effects stand out: the arrival of foreigners (76%), the privatization of natural resources (74%), increase in false expectations generated by the companies to the communities (70%), and different forms of violence (67%). In 79% of reported projects there was some type of violence (i.e., threats, intimidation, and forced displacement) in the resistance processes.

When asked about perceived environmental and cultural effects of extractives in Colombia, 82% of the surveyed jurisdictions in 2020 affirmed that there are no positive environmental effects of mining, a significant increase of 77 percentage points compared to 2013. The main positive environmental effects are watershed conservation (24%), reforestation (12%), and recovery plans for damaged areas (20%). The main negative environmental effects are pollution and/or deterioration of water sources (78%), soil deterioration (75%), and deforestation (78%). Prevalent cultural effects of extractives are being uprooted from territories (72%), a culture of waste (63%), and acculturation and loss of traditions (60%). Few projects had introduced programs for recovery of traditions of affected communities (17%).

Analyzing the Catholic Church's action on mining, development, and peace: the emergence of an eco-theology of peace

The Colombian Catholic Church, in compliance with its mission, works for reconciliation and unity, as well as to promote human dignity and the common good. Socially marginalized groups are the focus of the commitment to promote integral and sustainable development and work for justice and deeper reconciliation. The poor are the most affected by the conflicts caused by economic growth and social inequality, and the negative effects of mining on the environment. The church accompanies these communities, not only as a permanent presence in their midst but also to build on its history of struggling to build peace. Below we describe the implications of our study for the church's role in addressing mining issues in Colombia.

Strategies within the church

Although half of the jurisdictions partially agree that mining is necessary for the community, the Catholic Church is one of the leading organizations resisting mining. Leaders of community-based organizations and affected communities consider the Catholic Church a more trustworthy partner than academia or grassroots organizations. Of the many mining-related activities, church representatives are least likely to participate in political processes, but, when they do, the intensity of their participation is the highest. The next most intense levels of participation are teaching skills to local inhabitants and mediating between the government and the community. Clearly, the church in some jurisdictions is more active in accompaniment processes and could motivate other jurisdictions to do the same.

Our finding that mining operations have expanded in recent years to cover 90% of the territory challenges the Catholic community to consider how it might address this issue in a coordinated way in a larger number of local communities, as well as on a national and international level. The role of multinational corporations, foreign governments, and international institutions suggests the importance of coordinating with Catholic entities globally in addressing mining. At the local level, several strategies could be implemented, including (1) using the survey instrument as a pedagogical tool; (2) improving the space for interaction between government authorities and church leaders at the national level; and (3) providing training on extractive resources, mining impacts, accompaniment, and Catholic social teaching.

Our study helps address the lack of information on mining. While it might not be the role of the institutional Church to monitor such activities on an on-going basis, as is done in the Democratic Republic of the Congo, our study shows the need for Catholic and secular NGOs to develop or strengthen such monitoring as an essential first step in strengthening their capacity to engage effectively on mining issues. A communication portal on extractives at the jurisdiction level is necessary as it will provide local and national church leaders with vital information. This includes contact details of industry representatives and other mining actors, resources and data on extractives, a bank of documents on Catholic teaching related to the subject, and guidance on resistance processes and consultation or community engagement processes.

Strategies alongside the most vulnerable

The church's accompaniment involves recognizing the unity of community identity, culture, and territory. Mining-related conflicts rupture this unity. For example, the greatest victimizing event of the internal armed conflict in Colombia has been forced displacement – 7.5 million people have had to leave their territory. Mining modifies the lives of the local inhabitants, including their geographical environment and their customs. When those responsible for mining are unaware of or indifferent to such facts, new conflicts arise, existing conflicts are exacerbated, or latent conflicts intensify. Although an effective response must address the social and political structures of the country, the church is doing its best to help the urgent needs of the most vulnerable, as it does on a wide range of issues.

In order to reduce vulnerabilities and to prevent or reduce conflicts, the church must help communities take a constructive and proactive position towards their rights related to the environment and their territories. A multi-stakeholder dialogue requires training and accompaniment to build public spaces of dialogue and reciprocal listening. The main church activities in relation to extractives are (1) emotional, social, pastoral, and spiritual support; (2) presence at the base or directly serving the community affected by mining; and (3) advocating or working against the harmful effects of mining. In addition, the church trains community leaders and supports local peace initiatives that promote the collective construction of new relationships with the land and natural resources. The local church, Caritas, and the community can also collaborate in raising awareness of the situation for local, national, and international audiences. In addition, some pastoral work is directed towards government authorities and the private sector so that they effectively recognize the rights of the people and their obligation to comply with environmental protection provisions.

Communities in remote places and without employment options must deal with the dilemma of choosing between a system that guarantees daily sustenance and work in the short and medium term and preserving water sources and an environment conducive to agriculture and food security in the long term. A model of resource extraction that includes the well-being of the community and environment is possible to design and implement. It is an illusion to assume that there must be a trade-off between caring for the environment and mining-related development. Communities with mineral resources often ask themselves about their moral responsibility in caring for their environment. The church could help answer these questions. In particular, when communities are not consulted and face the challenge of caring for creation amidst violence by those who would harm the environment, pastoral care requires a pedagogy that helps the communities understand the theological meaning of creation, its universal destiny, and how to exercise their duty to protect the environment.

Anthropological foundation of ecology

We must not separate individual faith from collective care for creation, although for many people the former is not a prerequisite for the latter. An approach to ecology that stems from belief in God, the Creator, conceives the human person as the Imago Dei. Thus, a robust anthropology must be grounded in the value of human life. Respect for human life is a necessary condition for a coherent ecology.

In our study, we found that church representatives used similar principles of Catholic social thought to address the issues of extraction. Integral human development is used the most, followed by the common good and the promotion of peace. Surprisingly, stewardship of or care for creation were not noted in our survey. We also found that in 12% of cases respondents reported that they do not use any principles of the church's social doctrine. These results show the unfortunate separation between pastoral work and Catholic social teaching.

Integration of creation and salvation

The theology of salvation, a primary concern of the faithful, must be in dialogue with the theology of creation, which is neither well-known nor well-taught in many sectors of the church. A utilitarian approach to the goods of creation takes hold in the absence of a tangible sense of stewardship towards creation. Such a mentality endures when a faith that integrates the mystery of creation and the destiny of salvation is not cultivated. Integral and sustainable development and the principle of human dignity are in close relation to the call to care for creation (Francis 2015). Accompaniment for the Colombian Church needs to entail introducing these principles and responsible collaboration with the work of the Creator.

The National Social Pastoral aims to reconcile the focus on integral, solidaristic, and sustainable human development among church leaders with the principle of care for creation. The papal magisterium has advanced reflection on this throughout the last decades, especially recently with Pope Francis's encyclicals, *Evangelii Gaudium* and *Laudato Si'*, and his post-synodal apostolic exhortation, *Querida Amazonia*. The Latin American Episcopal Council has devoted much reflection to this topic and made calls for ecological awareness and care for creation (see CELAM 2018). In our study, the three most important resources of Catholic social teaching that are used in the accompaniment of communities in the jurisdictions in relation to extraction are *Laudato Si'*, the *Compendium of the Social Doctrine of the Church*, and the Bible.

Reconciliation and unity

Strengthening the relationship between peacebuilding and extractive activities allows a more constructive path for transforming conflict. Many cases in Colombia show the difficulties that people have in reestablishing neighborly relations and ensuring communal belonging after conflict subsides. It takes time to reconstruct an altered or destroyed social fabric. Pastoral action must consider that mining affects people's lives and puts them in very complex situations. Mining creates stress in personal lives and changes in the family structure and in the organization of communities. To speak about how to reconstruct the social fabric from a theological and pastoral perspective leads us to a deep reflection on the path traveled by the apostles and the nascent community after the resurrection and ascension of the Lord. The good was taken away and the better was coming with Pentecost. Those are the keys to accompany the community in the midst of bewilderment and numerous losses, including land, community relations, and its primary roots.

There is a permanent goal in pastoral work: to maintain hope. Hope is present in pastoral care as a deep, non-destructive relationship, one that is transformative and bears responsibility for the environment. In constructing peace in the context of environmental conflicts, which are often associated with mining, we have learned that we must not forget that hope is closely linked with harmony in the relationship between the inhabitants and the environment. This implies that working towards peace requires introducing a key ethical criterion: the care of creation. This element goes beyond the interaction between groups with conflicting or diverse interests, the intervention of the state, or the market. We cannot think of conflicts without recognizing that creation and its destiny are at stake. Not acknowledging creation or treating it in a simply instrumental way only deepens conflict and makes it much more complex in its different dimensions.

Peace requires a truly communitarian attitude. The church continues to promote open and transparent dialogue between different actors in society who are involved in socio-environmental conflicts. Experiences in Colombia show that when communities act as a whole and in a coordinated way, placing the common good above particular interests, they create new alternatives and perspectives to undertake the challenges faced in their territory. Pastoral experience shows that following this path helps to create links with government authorities and with other organizations interested in mining. The result is an integrated response to the problem, a community open to dialogue, which incorporates new perspectives and ways of seeing reality. Pope Francis teaches that unity prevails over conflict, but that requires that we carefully consider the impact on communities of conflict and alterations of their territory. By strengthening the cohesion of the community, recovering the sense of community, and constructing agreements that enable dialogue and advocacy with those responsible for political and social decisions, the "culture of encounter" is regained and strengthened.

Notes

1 This paper is based on a study conducted by Sandra Polanía-Reyes in collaboration with the Colombian Catholic bishops' National Social Pastoral and Caritas Colombia, directed by Msgr. Héctor Fabio Henao. We thank the Catholic Peacebuilding Network (CPN), based at the University of Notre Dame's Kroc Institute for International Peace Studies in the Keough School of Global Affairs, for financial support. The opinions expressed in this manuscript are the exclusive responsibility of the authors, and may not coincide with those of the National Social Pastoral, Caritas Colombia, or CPN.

2 For example, Instituto de Estudios para el Desarollo y la Paz (2019) reported that 566 killings occurred between January 2016 and January 2019 alone, and Cuartas's (2021) amicus brief for Human Rights Watch acknowledges that Colombia's Human Rights Ombudsperson's Office reports over 700 killings between 2016 and 2021.

3 See https://impactoindustriasextractivas.caritascolombiana.org/.

References

Agenda Propia. 2020. "Defensores de los cerros sagrados." January 2020. https://www.agendapropia.co/defensores-de-los-cerros-sagrados/#Inicio-lxms8KjhGG.

Alliance for Responsible Mining. 2014. "Update from Oro Verde and AMICHOCO." May 14, 2014. https://www.responsiblemines.org/en/2014/05/update-from-oro-verde-and-amichoco/.

Colombian Ministry of Foreign Affairs. 2021. "Distribución de Venezolanos en Colombia—Corte 31 de Enero de 2021." March 3, 2021. https://www.migracioncolombia.gov.co/infografias/distribucion-de-venezolanos-en-colombia-corte-31-de-enero-de-2021.

CELAM (Consejo Episcopal Latinoamericano). 2018. *Discípulos Misioneros: Custodios de Nuestro Hogar Común: Discernimiento a la Luz de la Encíclica* Laudato Si'. Bogotá: Consejo Episcopal Latinoamericano, CELAM.

Cuartas, José Fernando Reyes. 2021. "Amicus Brief on Killings of Human Rights Defenders in Colombia." Human Rights Watch. April 20, 2021. https://www.hrw.org/news/2021/04/20/amicus-brief-killings-human-rights-defenders-colombia#_ftn1.

Francis. 2015. *Laudato Si'*. https://www.vatican.va/content/francesco/en/encyclicals/documents/papa-francesco_20150524_enciclica-laudato-si.html.

Instituto de Estudios para el Desarollo y la Paz. 2019. "566 líderes sociales y defensores de derechos humanos han sido asesinados desde el 1 de enero de 2016—Al 10 de enero de 2019." January 11, 2019. http://www.indepaz.org.co/566-lideres-sociales-y-defensores-de-derechos-humanos-han-sido-asesinados-desde-el-2016-al-10-de-enero-de-2019/.

Peace Accords Matrix Barometer Initiative/Kroc Institute for International Peace Studies. 2021. *Executive Summary, The Colombian Final Agreement in the Era of COVID-19: Institutional and Citizen Ownership is Key to Implementation*. Notre Dame, IN and Bogotá, Colombia: Peace Accords Matrix Barometer Initiative/Kroc Institute for International Peace Studies/Keough School of Global Affairs. 10.7274/r0-cffp-qr07.

Polanía-Reyes, Sandra. 2021. "The Catholic Church and Extractive Activities in Post-Conflict Colombia: A Survey Follow-up to the Ecclesiastical Jurisdictions." Technical Report. Catholic Peacebuilding Network and Caritas Colombia (forthcoming).

Rettberg, Angelika, Carlo Nasi, Ralf J. Leiteritz, and Juan Diego Prieto Sanabria. 2020. *Different Resources, Different Conflicts? The Subnational Political Economy of Armed Conflict and Crime in Colombia*. Bogotá: Universidad de los Andes. https://www.jstor.org/stable/10.7440/j.ctv11vcd7n.

UN Human Rights Council. 2019. "Visit to Colombia: Report of the Special Rapporteur on the situation of human rights defenders." A/HRC/43/51/Add.1. December 26, 2019. https://undocs.org/en/A/HRC/43/51/Add.1.

4 The mining industry, conflict, and the Church's commitment in the Democratic Republic of the Congo

Rigobert Minani, SJ[1]

In 2019, the links between the extractive industry and the escalation of conflicts in the Democratic Republic of the Congo (DRC) were documented in a report to the UN Security Council (30–38). For over two decades, experts on the Great Lakes region, including diplomats, researchers, NGOs, and journalists, have identified the looting of mineral resources in the DRC as a structural cause of conflict (UN Security Council 2001; Jacquemont 2009; Cuvelier and Raeymaekers 2002). Since the outbreak of war in 1996, the plundering of natural and mineral resources in the DRC has fueled, escalated, and prolonged violence. The Church in the DRC has made a significant commitment to promoting good governance of these resources in order to help halt conflict and consolidate peace and stability.

Origin of the problem

The current problem did not start with the war of 1996. It goes back to the time of King Leopold II and the Berlin Division of colonial Africa in the late nineteenth century, as well as the turmoil of the postcolonial era. After independence in 1960, the country experienced significant sociopolitical difficulties. By 1975, the economy was failing, due to the fall in the price of copper, the oil crisis, the closure of the export route through the port of Lobito in Angola, and the hazardous economic policies of the Mobutu regime (Minani 2007, 502). These economic issues coincided with the Shaba War (March 1977–May 1978) and the massive departure of expatriate staff (Trinquier, Duchemin, and LeBailly 1963; Brassart 2018). The annual revenue of Gécamines, the state mining company, dropped from US$10 billion in 1974 to US$3.8 billion in 1990 (National Assembly of the Democratic Republic of the Congo 2006, 3).

On the eve of the DRC's series of wars starting in 1996, the country had been mired in a failed democratization process for at least seven years, its economy was at its lowest, and international cooperation had been suspended since May 1992. In short, the state was politically unstable and economically asphyxiated. These combined factors favored the outbreak of war, making possible the invasion of the country, the collapse of its security

DOI: 10.4324/9781003094272-4

apparatus, and a scramble to loot the DRC's mineral resources that involved more than nine African countries.

Privatization of the mining sector and escalation of the conflict

A few months before the war started, the World Bank and the International Monetary Fund came to the rescue of the Congolese economy. The country was asked to align with the World Bank's Structural Adjustment Policy. In 1995, under the aegis of these institutions, the country launched a major privatization of public enterprises, particularly the mining sector. According to a 2006 report from the DRC National Assembly, "The objective of this operation was twofold: pay down the country's debt by having foreign companies pay fees directly to international financial institutions, and replenish the state's coffers which had been drained by lax budget management" (5).

This operation allowed the wolf into the sheepfold. This was not the best time for this economic change. Internally, the Mobutu regime was at its end. Since 1992, opposition parties, churches, and civil society groups had asked Mobutu to leave power and the winds of democratization were shaking the country. The Mobutu regime did not, at this time, have sufficient credibility or legitimacy to execute a massive economic overhaul. With the end of the Cold War, international markets favored the mobility of capital on a global scale. The international financial institutions' economic reform measures were therefore very risky and poorly timed (Minani 2007, 503). And the privatization of mining companies in the DRC was fatal for its economy. The action amounted to a selloff of national wealth and it empowered illegal armed groups and other military actors. The period also coincided with the explosion of violence in the Great Lakes countries, with war in Burundi and genocide in Rwanda. The DRC experienced not only internal violence, but also spillover fighting from other countries near its borders. Mining companies raced to obtain mining titles sold at low prices because of war. And they also negotiated with the groups that had military control in the areas to be developed, giving the armed groups resources and power. From that moment on, control of mines became part of the war strategy.

The outbreak of conflict and the policies of the international financial institutions led the faltering government in Kinshasa to accelerate the slicing of its mining portfolio and the selloff of the nation's mining heritage. The government needed, on the one hand, to have funds to face the war, and, on the other hand, to build new alliances. Therefore, while the east of the country was occupied by rebels and foreign troops, the government accelerated the signing of agreements that sold off mining resources, sacrificing long-term sources of revenue and economic stability for needed cash-in-hand (Minani 2007, 505).

Mineral resources as fuel for conflicts in the DRC

Both the government in Kinshasa and the rebel group of Laurent Désiré Kabila engaged in haggling over mines with various actors, including criminals. Soon after the war began the government could no longer be a reliable partner for companies because it quickly lost control of a large part of its territory. The mining companies began to deal with the rebel leader Kabila in Goma. The rebels took advantage of these solicitations to strengthen themselves financially and militarily, and some of the contracts they signed with companies were for sites that had already been negotiated by Kinshasa, effectively redistributing revenue from the government to the rebels. This allowed Kabila to pay short-term bills, pay day-to-day expenses of his army, and fund his politico-administrative apparatus (Baracyetse 1999, 8). Helped by this money from mining, the support of neighboring countries, and international complicity, Kabila took power on May 17, 1997. Hardly a year later, on August 2, 1998, a new war broke out. Exacerbated by mining, this war became worse than the first one.

During the second war, rebel political alliances were reconfigured. The country was divided into five regions, each controlled by one faction.[2] Each group retreated to the mines in the areas it controlled to finance the group's war efforts. The rebel movements occupied mainly the eastern and northern DRC, while the government controlled the western and southern areas. That gave the government control of copper and cobalt mines in Katanga. For the rebel groups, the exploitation of gold in Ituri province, diamonds in Kisangani region, and gold and coltan in North and South Kivu became important.

The period also corresponded to a major spike in international demand for coltan, an important component in the personal electronics boom in the West. Coltan was exploited by, among others, armed groups, and marketed by neighboring countries, mainly Rwanda. It passed through refining plants in Europe, America, and Asia, and reached high-tech industries through a complex circuit (Cuvelier and Raeymaekers 2002, 9).

This rush for mining tipped the DRC into a criminalized economy. As I have written elsewhere, "The political weight of each rebel faction was proportional to the natural resources it controlled, according to its own rules of the game and its relationship with its allies" (Minani 2007, 507). Additionally, foreign countries involved in the conflicts, namely Rwanda and Uganda, were allowed to exploit the minerals in the DRC in exchange for support for rebel factions.

On April 1, 2003, the DRC concluded a peace agreement and put in place a transition government. Foreign troops were asked to withdraw, but neighboring countries, especially Rwanda and Uganda, strengthened their links with new armed groups in Kivu, Ituri, and elsewhere in eastern DRC in order to maintain control over the territory and access to natural resources. According to UN reports this was part of a long-planned strategy

(UN Security Council 2003, sec. 43–47). It escalated conflicts that have remained impervious to multiple peace initiatives. The death toll and the humanitarian crisis caused by this war, financed by the minerals in the DRC, mobilized a large number of international NGOs. Many Catholic actors in the DRC and around the world participated in several campaigns, with mixed results. None has succeeded in putting a final end to this scourge, but Congolese Catholics and their international partners have formed a powerful coalition to face these seemingly interminable problems.

Good governance of the mining sector and the Church's commitment

The Catholic Church in the DRC has always been at the forefront of social and political engagement, and the impact of extractive industries on conflicts has been a major concern since the early 2000s. As the Congolese bishops noted in 2007, their action on mining arises from the fact that extractive industries are a source of great wealth for some while being a source of great harm for many:

> Instead of contributing to the development of our country and benefiting our people, minerals, oil, and the forest have become causes of our misfortune. How do we take the fact that our fellow citizens, without consideration or compensation, are stripped of their land...? Is it permissible for Congolese workers to be treated without regard for their rights and human dignity? (§10)

The Jesuit Center of Study and Social Action (CEPAS)[3] took the lead on this issue. In 2006, it initiated a forum of civil society organizations to reflect on good governance of natural resources. This forum led to a campaign, initiated in March 2007 by DRC civil society groups and a coalition of international NGOs, calling for a revision of mining contracts extremely unfavorable to the DRC. A press release from the coalition stated: "The NGOs are calling on the Congolese Government and its international partners to renegotiate, revoke, or cancel the contracts inherited from the war and the transition in order to ensure that the Congolese people derive a fair share from the benefits of the exploitation of the country's mineral wealth" (A Fair Share for Congo/Une part équitable pour le Congo 2007). The campaign requested assurances of transparency, from the government and the World Bank, for past mining contracts. Specifically, it asked that the Congolese government establish a working group of independent international and Congolese experts to: review and revise all mining contracts signed during the war; establish an independent mechanism for monitoring the execution of contracts; and enshrine conditions for transparent and equitable management of mineral resources in the future (Minani 2007, 515).

To anticipate the results of such an exercise, CEPAS engaged mining evaluation experts, including former managers of Gécamines, to study the clauses of twelve contracts signed between the government and mining companies ("Republique Democratique du Congo revisitation des contrats miniers" 2007). They came to the conclusion that the review of the mining contracts was necessary so as to correct violations of the Mining Code, revive the economy, and improve the socioeconomic condition of Congolese citizens. This initiative laid a foundation for the government to revisit sixty-one contracts signed during the war. The Catholic bishops' conference of the DRC (CENCO) and CEPAS were invited to observe the process. At the end, the Ministry of Mines of the Democratic Republic of the Congo (2009) entrusted publication of the official government report to CEPAS. Unfortunately, when the time came to actually renegotiate contracts, it was done exclusively between the government and mining companies with no outside experts or observers. And because of confidentiality language included in the contracts, no details were made public.

Around this time, ecclesial leaders began focusing more on issues of mining. In 2007, the Assembly of Bishops of the Ecclesiastical Province of Lubumbashi, a group of pastors in Katanga, the main mining province of the DRC, wrote a pastoral letter that applied elements of the church's social doctrine to governance of extractive industries. This letter strongly condemned the fact that "the more investors ... invade Katanga, the more poverty, unemployment, and social problems increase" (§31). It affirmed the fact that minerals were neither benefiting the population nor the state. Workers were getting poorer, local economies were not strengthened, and artisanal miners were being arrested, tortured, and even killed. The more companies' profits grew, the more local communities suffered. It is against this background that the bishops of Katanga demanded that the Congolese state put the mining sector in order: "Congo is neither for sale nor on sale. ... The riches of our province must benefit our people ... [A]nd working conditions [must be] consistent with human dignity" (§34). The bishops also encouraged the government to diversify its economic development strategies by investing in agriculture, tourism, timber, and fishing in anticipation of the depletion of mineral resources.

Inspired by the work of CEPAS and the witness of the Katanga bishops, CENCO, in July 2007, decided to create an ad hoc episcopal commission in charge of natural resource governance:

> CENCO is committed, through its structures, to follow with particular attention the process of 'revision' of mining and forests contracts so that it can be done in the greatest transparency for the benefit of the Congolese people. It believes that the proper management of national resources will be a determining factor for the rebirth of the DRC. CENCO will create an ad hoc Episcopal Commission within its own

> body responsible for monitoring the issue of the exploitation of natural resources and demanding respect for human rights by companies working in this sector. (§10)

In 2013, the ad hoc group became a full commission of CENCO, the Episcopal Commission for Natural Resources (CERN). Its mission is to study all issues related to the exploitation of natural resources and propose to the church means of intervention to promote good governance of these resources and prevent conflicts.

CERN has played several roles. It acts at the level of decision-makers through advocacy to influence laws and policies related to the exploitation and management of natural resources in the DRC. It informs, trains, and raises awareness for citizens' control and management of natural resources, and advocates for the adoption of responsible economic, social, cultural, and environmental lifestyles. It works with local, national, regional, and international civil society organizations with similar objectives, as well as with other politico-administrative and economic bodies. CERN has established natural resource observatories in dioceses that document cases of human rights violations against communities and promote public awareness.

CENCO's deep engagement on the 2010 Dodd-Frank Wall Street Reform and Protection Act is a key example of the systematic, transnational approach the church has taken on mining. This US legislation included landmark provisions to ensure greater transparency in the supply chain of minerals. CENCO was centrally involved in the campaign to get Dodd-Frank passed. During the time when the legislation was taking shape, the president of CENCO was Bishop Nicolas Djomo Lola. His leadership and lobbying of Western governments were very influential in the law's formation (Carney 2014, 109). He testified before Congress in 2010 to support the law's passage, and again in 2012 during a hearing about the law's effects. At that 2012 hearing, he argued that the law should not be softened to make the regulatory process easier on businesses:

> We urge the U.S. business community to account for the gruesome social costs of the illicit mining as they calculate their costs for compliance with Section 1502. These calculations are not just cost estimates on a spreadsheet. There is a social balance sheet that places value on the lives that can be saved. We have full confidence in the goodwill of the Congress, the SEC [the US Securities and Exchange Commission], and the business sector to resist watering down SEC regulations through half measures that may save money, but cost lives. (US Congress 2012, 21)

Since Dodd-Frank's passage, and since the passage of similar regulations in Australia, Canada, and the EU, CENCO and CERN have continued to support transparency efforts. In 2017, when many believed that the United States might repeal the conflict mineral provisions in Dodd-Frank, CENCO's Henry Muhiya argued that the repeal would lead to new risks of conflict over minerals and would undermine local development programs begun as part of the DRC's efforts to comply with the law (Bagnetto 2017). In 2019, with financial support from CORDAID, CERN published a study on the impact of artisanal mining on local development. They then organized a workshop in 2020 with government and civil society representatives to examine how artisanal mining can be integrated into transparency regulation (ITIE-RDC Communication Unit n.d.). These independent, small-scale miners could be blocked from meaningful market participation if they are not enabled to navigate the regulatory frameworks made necessary by transparency laws. The recommendations emerging from the workshop included calling on the government to make sure technical assistance for transparency regulation is made available to these miners and that transparency frameworks are created that are specific to the artisanal sector. These kinds of initiatives led the Extractives Industry Transparency Initiative (EITI) (2019, 25) to highlight CERN as one of the leading Congolese organizations engaged in advancing EITI standards.

Conclusion

The experience of the DRC shows that a better understanding of mining's influence on conflict could help improve interventions to prevent and reduce conflict, and build lasting peace. The social doctrine of the church can help reveal the limits of the promises of the extractive industry (Minani 2017). Pope Francis (2015, §93) reminds us that "A type of development which did not respect and promote human rights—personal and social, economic and political, including the rights of nations and of peoples—would not be really worthy of man." He criticizes industries, such as mining, with short-sighted development models. Such development is at odds with the teaching of the church in the sense that it proposes a massive exploitation of resources without worrying about future generations. It aims at a rapid enrichment of investors and leaves local communities in poverty.

The extractive industry also has serious environmental consequences, including deforestation, land degradation, and contamination of air and water. The actors in the industry, including the government, armed groups, and multinational companies, also promote the expulsion of communities from their lands and contribute to massive violations of human rights. This situation in the DRC has contributed to violent conflicts, including an explosion of militias in Kivu today. The church's action for conflict transformation and peacebuilding in the mining sector is complex. It is, as Pope Francis (2020, §14) says, a fight against an economic instrument of death. The church needs to help the mining industry to better understand and care about the impact of its operations and the security and well-being of populations. The Catholic Church in the DRC has been actively engaged in many ways to try to help bring about this change.

Notes

1 Translated from French by Florence Silole.
2 West and Southeast of DRC: the Kinshasa government; North Kivu, South Kivu, Maniema and Sankuru regions: Rassemblement congolais pour la démocratie nationale-Goma; Equator region: Mouvement de liberation du Congo; Ituri region: Rassemblement congolais pour la démocratie nationale-Kisangani; Beni and Lubero region: Rassemblement congolais pour la démocratie nationale-K/ML.
3 See https://www.cepas.online.

References

Assembly of Bishops of the Ecclesiastical Province of Lubumbashi. 2007. "Que notre espoir ne soit jamais décu (Cf Ps, 70, 1)." March 3, 2007. https://www.congoforum.be/fr/2007/03/11-03-07-que-notre-espoir-ne-soit-jamais-du-message-de-lassemble-des-evques-du-katanga.

Bagnetto, Laura Angela. 2017. "Future of DR Congo Mining Could Be on Rocky Ground if US Dodd Frank 1502 Rolled Back." *RFI*, March 7, 2017. https://www.rfi.fr/en/africa/20170307-future-drc-mining-could-be-rocky-ground-if-us-dodd-frank-1502-rolled-back.

Baracyetse, Pierre. 1999. "L'enjeu géopolitique des Sociétés minières internationales en République démocratique du Congo (ex-Zaïre)." Press release for SOS Rwanda-Burundi Buzet (Belgium).

Brassart, Pierre. 2018. *Kolwezi 1978: Au Cœur des opérations françaises et Belge au Zaïre*. Paris: Mardaga.

Carney, J.J. 2014. "'The Bishop is Governor Here': Bishop Nicolas Djomo and Catholic Leadership in the Democratic Republic of the Congo." In *Leadership in Postcolonial Africa: Trends Transformed by Independence*, edited by Baba G. Jallow, 97–122. New York: Palgrave Macmillan.

CEPAS. 2006. *Bonne gouvernance et ressources naturelles et minières en République démocratique du Congo*. Kinshasa: Éditions CEPAS.

Cuvelier, Jeroen, and Tim Raeymaekers. 2002. "Supporting the War Economy in the DRC: European Companies and the Coltan Trade." IPIS Report. January 16, 2002. https://ipisresearch.be/publication/supporting-the-war-economy-in-the-drc-european-companies-and-the-coltan-trade.

EITI International Secretariat. 2019. *Validation of the Democratic Republic of Congo: Report on Initial Data Collection and Stakeholder Consultation*. April 15, 2019. https://eiti.org/files/documents/drc_2018_validation_initial_assessment_final_april_2019.pdf.

A Fair Share for Congo/Une part équitable pour le Congo. 2007. "International Appeal for a Revision of the DRC's Mining Contracts." Press Release. March 13, 2007. http://www.raid-uk.org/sites/default/files/pr-contracts-campaign.pdf.

Francis. 2015. *Laudato Si'*. https://www.vatican.va/content/francesco/en/encyclicals/documents/papa-francesco_20150524_enciclica-laudato-si.html.

Francis. 2020. *Querida Amazonia*. http://www.vatican.va/content/francesco/en/apost_exhortations/documents/papa-francesco_esortazione-ap_20200202_querida-amazonia.html.

ITIE-RDC Communication Unit. n.d. "Cern-Cenco: plaidoyer pour l'intégration du secteur minier artisanal à l'ITIE et au financement du ST par le Gouvernement." ITIE-RDC. Accessed April 22, 2021. https://www.itierdc.net/2020/09/24/cern-cenco-plaidoyer-pour-lintegration-du-secteur-minier-artisanal-a-litie-et-au-financement-du-st-par-le-gouvernement.

Jacquemont, Pierre. 2009. "La sortie de crise dans l'Est du Congo et les perspectives de la Coopération Régionale." *Mondes en Developpement* 37, no. 147: 93–108.

Minani, Rigobert. 2007. "La problématique des ressources minières en RDC: Etat de lieu et perspectives." *Congo-Afrique* 417: 501–518.

Minani, Rigobert, ed. 2017. *Unmet Promises of Extractive Industries in Africa*. Nairobi: Paulines Africa.

Ministry of Mines of the Democratic Republic of the Congo. 2009. *Commission de Revisitation des contrats miniers: Rapport des travaux*. 2 vols. Kinshasa: Éditions CEPAS.

National Assembly of the Democratic Republic of the Congo. 2006. *Commission spéciale chargée de l'examen de la validité des conventions à caractère économique et financier conclues pendant les guerres, 1996–1997 et de 1998*. Work report, part 1.

National Episcopal Conference of the Congo. 2007. "'A vin nouveau, outres neuves' (mc 2, 22): Ne pas décevoir les attentes de la nation." http://www.virunganews.com/a-vin-nouveau-outres-neuves-mc-2–22-ne-pas-decevoir-les-attentes-de-la-nation.

"Republique Democratique du Congo revisitation des contrats miniers: rapport du groupe d'experts du forum de la société civile sur 12 contrats miniers." 2007. http://pratclif.com/mines/kamoto/revisionContrats_miniers_Rapport_definitif.pdf.

Trinquier, Colonel [Roger], Jacques Duchemin, and Jacques LeBailly. 1963. *Notre guerre au Katanga*. Paris: La Pensée Moderne.

UN Security Council. 2001. *Democratic Republic of Congo Asserts Right to Seek Help in Defense of Natural Wealth Against Outside Forces*. Press Release SC/7241. December 14, 2001. https://www.un.org/press/en/2001/sc7241.doc.htm.

UN Security Council. 2003. *Letter dated 15 October 2003 from the Chairman of the Panel of Experts on the Illegal Exploitation of Natural Resources and Other Forms of Wealth of the Democratic Republic of the Congo Addressed to the Secretary-General*. S/2003/1027. October, 23, 2003. https://undocs.org/S/2003/1027.

UN Security Council. 2019. *Final Report from the Group of Experts on the Democratic Republic of the Congo*. S/2019/469. June 7, 2019. https://undocs.org/S/2019/469.

US Congress. 2012. House. Subcommittee on International Monetary Policy and Trade of the Committee on Financial Services. *The Costs and Consequences of Dodd-Frank Section 1052: Impacts on America and the Congo*.112th Congress, May 10, 2012. Statement of The Most Reverend Nicolas Djomo Lola, Bishop of Tshumbe, Democratic Republic of the Congo and President of the National Episcopal Conference of the Congo. https://www.govinfo.gov/content/pkg/CHRG-112hhrg75730/html/CHRG-112hhrg75730.htm.

5 Mining in the Philippines: a Catholic peacebuilding approach

Karl M. Gaspar, CSsR

In the Philippines, as in other countries examined in this book, mining is intricately connected to issues of peace, human rights, development, and ecology. Since Vatican II and especially since its role in the People Power movement that brought down the Marcos regime, the Catholic Church in the Philippines has been an influential force for social justice and peace. This essay connects the church's history of engagement on these broader issues over the past five decades to its more recent engagement on mining.

Conflicts in the Philippines

Since the Philippines gained independence in 1946, its situation has always been volatile, especially on the southernmost island of Mindanao. In the 1950s and 1960s, government forces crushed armed bands of Islamic Moros (see Constantino and Constantino 1978; Diokno 1987). A more determined armed resistance arose with the founding of the Moro National Liberation Front (MNLF), a separatist movement which began to recruit rebels in 1969 (Gaspar, Lapad, and Maravillas 2003, 29–30; see also Vitug and Gloria 2000). Violent encounters between MNLF and government forces took place across parts of Mindanao from 1972–75. In 1976, Libyan leader Muammar Gaddafi brokered a short-lived cessation of hostilities between the government and the MNLF. The signing of this agreement brought about a serious rift in MNLF leadership, leading to the formation of a breakaway group that established the Moro Islamic Liberation Front (MILF) in 1984.

In January 1987, the MNLF accepted the Philippine government's offer of semi-autonomy for the regions in dispute, leading to the establishment of the Autonomous Region of Muslim Mindanao on November 6, 1990. The MILF, however, refused to participate. As government administration changed from one president to another, peace talks were initiated but never succeeded. It was during the presidency of Rodrigo Duterte that the government and the MILF finally agreed on the Bangsamoro Organic Law, signed by President Duterte on July 26, 2018, and approved by plebiscite in early 2019. Consequently, the Bangsamoro Transitory

DOI: 10.4324/9781003094272-5

Parliament was established and is currently laying down the groundwork for a governance system.

However, it was not only the Moro rebellions that the state has had to deal with since the early 1970s. Whereas these only involved territories in the Muslim areas in Mindanao, the revolution that has been waged by the Communist Party of the Philippines (CPP)—with its military arm, the New People's Army (NPA)—operated across the country (see Sison 2013; Reid 2000). Fighting a guerrilla war against the state, the CPP-NPA remains an underground operation aimed at overthrowing the Philippine government, implementing a radical land-reform program, promoting a national economic policy, and ending US imperialist influence over its former colony. It has survived six presidents and remains a force to reckon with. In the wake of the People Power uprising in 1986, which installed Cory Aquino as successor to Ferdinand Marcos, the first peace talks between the state and the CPP-NPA were convened. They collapsed owing to the state military's belligerence. The talks were resurrected by other administrations, but would always begin with high hopes and end with bitter frustration on both sides (see Bolasco 2019). Duterte has expressed his interest in resuming talks, but only time will tell if he can succeed in finally ending this insurgency that has lasted more than fifty years.

As these rebellions have raged across the country, the civilian population has suffered considerable collateral damage. Farmers, fisherfolk, and indigenous peoples have been the most victimized by the consequences of military operations, which have often led to human rights violations (ABS-CBN Investigative and Research Group 2018).

Violent conflict, human rights violations, and mining

Holden and Jacobson (2007) lay out how mining worsens and prolongs violent conflict in the Philippines. First, mining companies and armed groups are frequently at odds simply because mining is most common in regions that have the preferred terrain for guerrilla warfare; this is especially true in Mindanao (485, 487). But that tension is only a starting point. Territorial rivalry makes mining companies clear targets for extortion. Holden and Jacobson cite one example of a company that was found to have paid over US$1.7 million to several different armed groups (487). This funding increases the capacity for guerrilla groups to wage war, while at the same time, robust fear and distrust of mining companies throughout the Philippines, especially in rural areas that depend heavily on environmental resources damaged by mining operations, gives armed groups fertile ground for recruitment (489). Further, mining companies are tangible symbols of the globalized economic system despised by the ideology of armed militants, making them appealing targets for attacks and furthering their usefulness as foils to enhance recruitment (489).

Given these risks, the typical response at mining sites has been increased militarization. National armed forces and private security have been quick to cast all opposition to mining as terrorism, or as in league with armed insurgents, leading armed security and military forces around mining operations to be draconian (490–92).

With this sort of pattern, human rights abuses have been common. Since the passage of the 1995 Mining Act, the media, indigenous persons organizations, and other civil society organizations (CSOs), many of which are faith-based, have reported violence and human rights violations, especially when mining projects face aggressive resistance. These have included reports on "beatings, maiming, intimidation, torture, extra-judicial killings or summary executions, abductions, and enforced disappearances in what is widely viewed as state-sponsored or state-protected terrorism" (Goodland and Wicks 2008, 44). Human Rights Watch identified anti-mining activists as victimized along with other groups active in the resistance movement, including political and student activists, journalists, and church workers (Shepherd and Neumann 2007, 2, 28).

Indigenous communities in particular have had a sordid history with human rights violations related to mining. In 2001, a report detailed how mining operations were violating the rights of the Subanens in the Zamboanga area of Mindanao (Sanz 2001). The report was issued by researchers who conducted a human rights impact assessment of the TVI Resource Development's Canatuan mine. It looked into six core human rights: self-determination, security, adequate standard of living, adequate housing, work, and education. It indicated that the mine had a negative impact on the right to self-determination and on local self-governance. The operation was also found to have forced evictions and increased militarization of the area around the mine, and to have "had a negative impact on the ability of the Subanon [sic.] to enjoy the human right to security and the human right to housing. Mining activity appears to have increased the levels of sediment and metals in some local waterways, threatening the human right to an adequate standard of living" (53).

A final point that Holden and Jacobson (2007) make about mining and conflict in the Philippines is that mining economically marginalizes large scores of the Philippine population. Mining has a poor record of job creation and high risk for environmental costs that inhibit people from engaging in other livelihoods, factors that leave poor people even poorer and disenfranchised people even more disenfranchised (494–96). This creates greater social instability that exacerbates existing conflict factors. Combine this economic problem with rampant corruption in the mining sector (492–93) and the militarization described above that increases tensions, introduces human rights violations, and enriches militants while strengthening their recruitment platforms, and mining and conflict become locked in an escalating spiral.

The Church's response to conflict, human rights, and environmental concerns

The engagement of the Church in the Philippines on mining is best understood in the context of its wider efforts to address issues of peace, justice, and ecology from the Marcos era to the present. Mainly through the Catholic Bishops' Conference of the Philippines (CBCP) and the Mindanao-Sulu Pastoral Conference (MSPC), the church took a strong prophetic stance to denounce human rights violations during the Marcos regime.

Between 1977 and 1979, a new challenge confronted the bishops. On one hand, seven years of Marcos's dictatorial rule further worsened as poverty increased, foreign debt accumulated, Imelda Marcos's lifestyle became more scandalous, corruption in the bureaucracy now involved most of Marcos's cronies who monopolized control over government corporations, and human rights violations skyrocketed as more suspected "subversives" were arrested or assassinated. At the same time, the underground communist movement had become quite popular. Reacting to this situation, the CBCP issued the 1979 pastoral letter "An Exhortation Against Violence." They critiqued violent reactions to the nation's problems, and described this situation as follows:

> The daily newspapers are filled with accounts of military encounters, of ambushes and killings, not to mention the more ordinary crimes, the scandals of corruption and other forms of violence. Revolutionary groups are increasing in number and escalating the spirit (and in some cases the praxis) of violence, especially among the urban and rural workers, intellectuals and students, and among a number of dedicated and concerned Christians, including some clerics and religious. (n.p.)

In clear terms they stated that, "In the world where violence all too frequently maims and destroys the personal and sacred lives of men, we must uphold the violence of love and the peace of Christ over hate and destructive violence ... The option of non-violence must be respected as one Christian option, as a Christian pattern of action" (n.p.). The statement ends with an exhortation towards justice and peace. It was no surprise that such phrases would appear in the bishops' statement, as they were worried that some priests, religious, and seminarians were becoming more sympathetic to the communist movement.

In February 1983, the CBCP (1983a) issued a new pastoral letter reacting to the military's arrest of priests and religious and lay church workers, many of whom became political prisoners. They denounced such human rights violations and cited the reasons why they were supporting the struggles of the poor, deprived, and oppressed, pointing back to the 1971 Synod on Justice. Six months later, on August 21, on his return to the Philippines from exile in the United States, Senator Benigno "Ninoy"

Aquino was assassinated on the Manila airport tarmac. With his martyrdom, the protest movement grew by leaps and bounds. However, such a movement only provoked more repression, prompting the bishops to issue a statement on reconciliation. They acknowledged that "many events have pushed our country closer to the brink of chaos and anarchy," and reiterated "the need for reconciliation as an alternative to the continuance of present injustice and violence which would put one against another in a bloodbath of revolution where the Gospel ethic of love would undeniably be sacrificed." Once more, they highlighted the path of "the non-violence of Christ ... as the only acceptable answer," as it is "a way of working constantly, strenuously for justice that refuses adamantly to destroy life for the cause of justice itself" (CBCP 1983b).

Another church body that was a major force for peacebuilding was the MSPC, constituted in 1971 by the dioceses of Mindanao-Sulu and meeting every three years. In the first two conferences, the delegates became more vocal in their opposition to martial law, which was declared in 1972 (see Gaspar 1977). At the MSPC in 1984, they demanded that Marcos lift martial law and return to the country's democratic tradition. However, martial rule only worsened. Instead of bringing about peace and order, there was more dislocation of communities given the twin evils of incursions into the countryside by corporations and the ensuing militarization that usually led to abuses committed by the armed and paramilitary forces providing security.

While the bishops' conference had not spoken about the Moro rebellions and the impact of violence on the citizenry, the MSPC conferences since the beginning expressed their concern for the plight of the Moro and the escalation of armed encounters leading to mass evacuations and human rights violations. They also discussed the impact of martial rule on the lives of indigenous peoples, whose situation could only be reported because of the presence of missionaries in their isolated territories. The chair of the conference, Bishop Bienvenido Tudtud, proposed beginning an interfaith dialogue with the Muslims which led to a program known as Duyog Ramadhan, a community-based activity where Muslims and Christians came together for a dialogue of life and faith. It also led to the establishment of the Bishops-Ulama Conference, a forum bringing together Christian and Muslim leaders which became instrumental in peacebuilding efforts. Meanwhile, the missionary congregations engaged in advocacy work among indigenous communities and began to come up with collaborative efforts to help protect them from the abuses of the martial law regime as well as to empower them to stand firm on their rights.

In the midst of these peacebuilding efforts, the mining and logging industries attracted attention from some church groups, as these were promoted as money-making ventures for Marcos and his cronies. Logging firms were owned primarily by Filipino oligarchs, a number of whom had run for political office so they could take advantage of their position in government to protect their interests. The few Filipino mining companies—again, owned by oligarchs who also owned other businesses, especially in manufacturing and trading—partnered with Western firms to mine gold, silver, and copper in various parts of the country. During the Marcos regime, however, no one had yet heard of global warming and climate change and how deforestation and mining contribute to this ecological crisis. And as many other justice and peace issues were more urgent, there were not many actions to protest the expansion of mining and logging.

The shift to ecological advocacy

From the time of the American occupation to the contemporary period, there has been no let-up in deforestation (see Forest Management Bureau 2009). But for a long while massive deforestation and the expansion of agribusiness plantations were not a major concern of the Church in Mindanao. The first stirrings that would lead to a nascent church movement engaged in ecological issues arose in the late 1980s in two areas: San Fernando, Bukidnon, and Midsalip, Zamboanga del Sur. This pastoral engagement was provoked by peasants' concerns about land-grabbing and the impact of logging in limiting water that could be tapped for their irrigated rice fields. The fact that the peasant communities were organized in vibrant base ecclesial communities and had the support of their missionaries made possible anti-logging mass actions that succeeded in terminating logging operations in the area.

These grassroots-based ecological actions influenced the CBCP (1988) to issue the pastoral letter "What is Happening to Our Beautiful Land?" As the mining problem had not yet become urgent, the statement hardly referred to this issue. However, it exhorted the government "not to pursue short term economic gains at the expense of long-term ecological damage" (n.p.). In a context when 29 out of the 30 million hectares of the country's primary forests had been destroyed, the bishops claimed that awareness of the relationship of people to the environment had continued to grow and that destroying the forests was sinful. They posited that "As people of the covenant we are called to protect endangered ecosystems like the forests, mangroves and coral reefs and to establish just human communities in our land" (n.p.).

Henceforth, the concern for ecological advocacy within the church began to slowly expand from diocese to diocese. In many cases, civil society organizations assisted them. One that was very active in the Davao area was the Kinaiyahan Foundation, and later the Interfacing Development Initiatives for Sustainability (IDIS). Various church groups supported IDIS's unfortunately unsuccessful campaign to end aerial spraying in banana plantations.

Church action and advocacy around mining

Mining became a focus of the Catholic community's ecological concern in the mid-1990s. Mining, especially open pit mining, became contentious following the Congress's approval of the Philippine Mining Act in 1995, which "liberalized the legal framework for mining, making it far less restrictive," as it allowed "greater foreign ownership, full repatriation of profits, tax breaks and tax holidays for five to ten years, and lower duties and tariffs" (Goodland and Wicks 2008, 23). Consequently, there arose a strong interest on the part of mining firms to expand their present operations and/or explore new territories where significant deposits of mineral resources had been identified. As soon as the Mining Act was passed in 1995, there was a huge rush of applications. Under the new law, two different licenses existed: Mineral Production Sharing Agreements (MPSAs), which limited foreign ownership of a project to 40%, and Financial and Technical Assistance Agreements (FTAAs), which allowed full foreign ownership. Before the major push toward expanding the area of mining coverage, there were just a few existing MPSAs, with only two existing FTAAs that covered 447,308.26 hectares. After the passage of the Mining Act, 54 more FTAA applications targeting 2,350,643.34 hectares were approved (Goodland and Wicks 2008, 25).

But the Act did not go uncontested. A few quarters within the government bureaucracy and civil society, including the La Bugal-Blaan Tribal Association of Mindanao, claimed the Act violated the 1987 Constitution's provision that "all lands of the public domain, waters, minerals, coal, petroleum and other mineral oils, all sources of potential energy, fisheries, forests, timber, wildlife, flora and fauna and other natural resources are owned by the state." Citing the devastating effects of some of the mining operations, such as spillages of mine tailings, the CBCP (1998, n.p.) argued that "the adverse social impact on the affected communities, especially on our indigenous brothers and sisters, far outweighs the gains promised by large-scale mining corporations." They then proposed that the government repeal the Act and recall all approved FTAAs and mineral agreements and disapprove the pending ones. Owing to the pressure from CSOs and the church, in 2004, the Supreme Court ruled the Mining Act was unconstitutional.

The state did not give up so easily and used all its power to pressure the Supreme Court to reverse their ruling. Ten months later, the Supreme Court changed its mind. President Macapagal-Arroyo immediately issued an executive order known as the National Policy Agenda on Revitalizing Mining. This provoked more protests as this decision impacted various stakeholders, including municipal and provincial government authorities, the church, and other CSOs, who all continued to pressure the Supreme Court to revert to its original decision.

Once more in 2006, the CBCP expressed its support for the many petitions against mining operations and re-affirmed its stand for the repeal of the Mining Act of 1995. The bishops stated that allowing the interests of big mining corporations to prevail over people's right to preserve natural resources necessary for their livelihoods was tantamount to violating their right to life as well as threatening their health and environmental safety. All kinds of mining destroyed the environment while depriving the people of the land they should be cultivating for their subsistence. Especially in relation to indigenous peoples, the bishops made it clear they were against any mining firm that would dislocate them from their ancestral domain.

In 2008, Leila de Lima was appointed chairperson of the country's Commission on Human Rights. She was aggressive about pursuing cases of human rights violations, and CSOs and church personnel could seek her assistance in following-up reports of violence. One such case took place in the village of Kasibu in the province of Nueva Viscaya on November 5–6, 2009. This location was where the mining operations of Oceana Gold Philippines, Inc. took place. The local parish supported de Lima's investigative campaign. It was a breakthrough case as an agency in government was seen as taking a strong position against human rights violations and giving peacebuilders and rights defenders an ally.

In the next few years, the CBCP continued its advocacy against mining. A July 2010 letter from Nerio Odchimar, Bishop of Tandag, to President Benigno Aquino III reiterated the bishops' decade-old request that the government put a stop to large-scale mining "since this not only permanently damages the delicate balance of the natural environment, but it also makes our small farmers, fisher folks and IPs [indigenous persons] suffer." In 2010, the Department of Environment and Natural Resources (DENR) issued Administrative Order No. 2010–21. This order formally amended the provisions of the 1995 Mining Act, upholding the principle that public domain lands, including those with natural resources like minerals, are owned by the state, and making space for the rights of indigenous persons in accordance with the 1997 Indigenous Peoples Rights Act.

The 2010 DENR order did not immediately resolve all disputes and tensions, as despite its safeguards, the liberalization of the industry continued to usher a massive expansion of mining exploration and operation; thus, the church's advocacy against mining persisted. The Association of Major Religious Superiors in the Philippines (2013) issued a joint statement on mining and greater solidarity with the poor. Appealing to their roles as stewards of the environment and as shepherds to indigenous communities, the religious superiors continued to lobby for policies protecting our common home against environmentally destructive activities. The visit of Pope Francis to Manila in January 2015 provided an opportunity for the CBCP's Secretariat for Social Action to issue a joint letter co-signed by Alyansa Tigil Mina (2015), an advocacy network for mining justice. The letter referred to a message of Pope Francis during a Day of Reflection between the Vatican and mining executives that reminded everyone that "the great challenge of business leaders is to create a harmony of interests, involving investors, managers, workers, their families, the future of their children, the preservation of the environment on both a regional and international scale, and a contribution to world peace" (n.p.).

Claiming the need to add to the gross national product while providing employment and increasing government revenues, administrations since President Macapagal-Arroyo have encouraged more investment in mining. At the start of his term in 2016, President Duterte appointed a secretary to the DENR who was staunchly anti-mining. But pressure from both the mining oligarchs and their supporters in Congress prevented confirmation of her appointment, and that ended this initial attempt to limit mining. Since then, Duterte has fully backed the expansion of mining, providing the same rationale as previous administrations. It is in Mindanao where most mining explorations are being pursued, including the new Bangsamoro areas and areas where indigenous communities like the Lumad and the Subanen reside.

In 2019, the bishops issued a statement supporting all those who continue to oppose mining. In the statement, they once more called the faithful's attention to the threat of extractive mining operations and reiterated that:

> Social justice is not served when only the few mining companies, many of which are also owned by political leaders, reap the benefits from mineral extraction. The rural poor remain poor as mining only contributes less than one percent to our GDP, employs less than 0.4% of our labor force, and directly threatens agriculture, forestry, watersheds and fisheries resources that are essential for the survival of the rural poor. (CBCP 2019, n.p.)

One specific type of mining this statement addressed was coal. The state has aggressively promoted coal because of increased reliance on coal-fired power plants and a desire to reduce reliance on foreign coal to fuel them. Across the country, these plants have sprouted in various places. The CBCP statement observed that by the end of 2020 there were expected to be over fifty operational coal plants and 186 coal-mining projects. Many are located in areas that have rich ecosystems and are mostly inhabited by indigenous peoples. The statement affirmed scientific evidence of coal causing climate degradation, and stated that "Coal projects also further exacerbate the vulnerability of impoverished host communities in the Philippines already struggling to cope with the effects of the worsening climate" (n.p.). Consequently, according to the bishops, coastal and agriculture-reliant communities have experienced losing their livelihoods because of land conversion and have experienced adverse health impacts due to pollution as toxic substances and heavy metals are released into the air.

Local government officials and indigenous leaders in all these areas are being wooed by mining firms that desire to begin operations soon. Where they have resisted, the people have been subjected to harassment and human rights violations. One area where this has been taking place is between the Diocese of Marbel and the Archdiocese of Cotabato, where coal deposits were recently discovered. The local social action centers opposed the opening of such operations and supported the indigenous communities who took a stand to oppose the coal mining.

Another major development that spurred strong church response occurred in 2021. President Duterte's EO 130 amended President Aquino III's EO 79 from 2012 to lift a moratorium on mineral contracts in protected areas, prime agricultural lands, tourist areas, and other critical locations. Caritas Philippines and other church actors strongly opposed Duterte's decision, contending it would harm the environment while mainly benefiting business interests, not the poor. Nine provinces in Mindanao, including Moro and Lumad areas, would be impacted by lifting the moratorium (Gomes 2021). The church's ability to respond to Duterte's decision was enhanced by Eco-Convergence, a church-civil society partnership to promote *Laudato Si'*. Eco-Convergence Hubs throughout the Philippines that track data and develop local expertise in monitoring mining operations enabled church leaders to highlight the likely impact on their communities of the lifting of the moratorium.

This recent affair points to one very important thing local churches will need if they are to sustain their advocacy against mining: data on how the mining industry is playing out in their specific territories. Information is not difficult to obtain, as it is publicly available online and there are media institutions interested in filing reports on mining explorations. By knowing exactly what sort of mining is operating or is planned in their respective dioceses, the local churches would be able to provide the data to their constituencies that is needed to formulate effective advocacy.

For example, one of the most vibrant forms of church action with regard to mining has been grassroots efforts to create opportunities for alternative economic activities in areas marked for mining development. In an interview with Holden and Jacobson (2007, 496), I discussed the widespread, parish-level efforts of base ecclesial communities in Mindanao to foster bottom-up development with activities like organic farming, microfinance, and handicrafts. These sorts of localized development projects, supported by local churches and encouraged in seminary training, are small but powerful vehicles for resisting the dominant power of the globalized mining industry. And it is essential that leaders are able to access accurate and up-to-date data on what the government and mining companies are doing and planning to do in their areas, as well as what impacts can be expected.

Conclusion

Since the country gained independence, social unrest has characterized Philippine society. Over the decades, a robust matrix of civil society work for peace has evolved. This has included labor organizing, youth movements, advocacy groups, and the influential People Power Revolution of the 1980s. These efforts came to also include conscientization programs so that the poor and disadvantaged would be made aware of the root causes of their marginalization. The formation of leaders assured that the empowerment of the poor would be sustained as the struggles would take time, and the mobilization of resources was also assured.

Another decades-long problem has been armed insurgencies. Unrest and violent conflict have been constants. As with other areas of social unrest, civil society organizations have also been deeply embedded in peacebuilding work in affected conflict zones and have made real progress. The various agencies operating under the Catholic Church, from episcopal commissions to diocesan pastoral centers to religious orders, have been influenced by and helped shape these efforts for justice and peace. The Church in the Philippines has been one of the most active and effective peacebuilding organizations in the country, and remains engaged on a range of issues, including ones related to ongoing strife with the government as well as with armed groups. Some of the factors that have made Catholic engagement effective in reducing violent conflicts and creating conditions for a just and sustainable peace include the following: strong social and symbolic capital; the wise use of a prophetic role that can influence a wide range of the population, including state officials as well as those holding corporate power; and vibrant base ecclesial communities.

Today, this peacebuilding capacity is being exercised in areas where there is conflict associated with mining explorations and operations. Mining has become a focal point of the government's economic development plans, and aggressive pushes to increase mining output, especially in the southern region of Mindanao, have introduced waves of human rights violations, violence, and new and worsened vulnerabilities for communities already dealing with the effects of decades of conflict. Indigenous peoples have been particularly impacted. These mining conflicts interweave environmental questions, human rights issues, social cohesion and conflict transformation challenges, poverty and economic marginalization, and indigenous rights. In this light, the church's experience with varied peacebuilding activities across levels of society, with developed sensitivity to ecological concerns, offers important resources to respond to these problems.

References

ABS-CBN Investigative and Research Group. 2018. "By the Numbers: Human Rights Violations During Marcos' Rule." *ABS-CBN News*, September 21, 2018.

https://news.abs-cbn.com/focus/09/21/18/by-the-numbers-human-rights-violations-during-marcos-rule.

Association of Major Religious Superiors in the Philippines. 2013. AMRSP Convention Statement. *AMRSP Joint Biennial Convention*, Davao City, July 1–5, 2013. http://www.pddm.ph/index.php/news/news-and-updates/1647-amrsp-convention-statement.

Bolasco, Maria Karina Africa. 2019. "The GRP-NDFP Peace Talks: Tactical Discontinuities in a Shared Narrative." *Kyoto Review of Southeast Asia, Trendsetters* 26. https://kyotoreview.org/trendsetters/grp-ndfp-peace-talks.

CBCP (Catholic Bishops' Conference of the Philippines). 1979. "Exhortation Against Violence." https://cbcponline.net/exhortation-against-violence-a-joint-pastoral-letter-of-the-philippine-hierarchy.

CBCP (Catholic Bishops' Conference of the Philippines). 1983a. "A Dialogue for Peace." https://cbcponline.net/a-dialogue-for-peace-joint-pastoral-letter-catholic-bishops-conference-of-the-philippines.

CBCP (Catholic Bishops' Conference of the Philippines). 1983b. "Reconciliation Today." https://cbcponline.net/reconciliation-today.

CBCP (Catholic Bishops' Conference of the Philippines). 1988. "What is Happening to our Beautiful Land?" https://cbcponline.net/what-is-happening-to-our-beautiful-land.

CBCP (Catholic Bishops' Conference of the Philippines). 1998. "A Statement of Concern on the Mining Act of 1995." https://cbcponline.net/a-statement-of-concern-on-the-mining-act-of-1995.

CBCP (Catholic Bishops' Conference of the Philippines). 2006. "Mining Issues and Concerns." https://cbcponline.net/a-statement-on-mining-issues-and-concerns-do-not-defile-the-land-where-you-live-and-where-i-dwell-num-3534/.

CBCP (Catholic Bishops' Conference of the Philippines). 2019. "An Urgent Call for Ecological Conversion, Hope in the Face of Climate Emergency." https://cbcpnews.net/cbcpnews/an-urgent-call-for-ecological-conversion-hope-in-the-face-of-climate-emergency.

Catholic Bishops' Conference of the Philippines National Secretariat for Social Action/Caritas Philippines and Alyansa Tigil Mina. 2015. "Joint Statement on the Mining Issue During the Visit of Pope Francis in Manila and Tacloban in the Philippines: 15–19 January 2015." https://hronlineph.com/tag/catholic-bishops-conference-of-the-philippines.

Constantino, Renato, and Letizia R. Constantino. 1978. *The Philippines: The Continuing Past*. Quezon City: The Foundation for Nationalist Studies.

Diokno, Jose W. 1987. *A Nation for Our Children: Human Rights, Nationalism, Sovereignty: Selected Writings of Jose W. Diokno*. Quezon City: Claretian Publications.

Forest Management Bureau. 2009. "Philippines Forestry Outlook Study." Working Paper No. APFSOS II/WP/2009/10, Asia-Pacific Forestry Sector Outlook Study II. Food and Agriculture Organization of the United Nations Regional Office for Asia and the Pacific. http://www.fao.org/3/am255e/am255e00.pdf.

Gaspar, Karl. 1977. "The Growth and Development of MSPC." In *What is MSPC?*, 14–17. Davao City: MSPC Publications.

Gaspar, Karl, Elpidio A. Lapad, and Ailynne J. Maravillas. 2003. *Mapagpakamalinawon: A Reader for the Mindanawon Peace Advocate*. Davao City: Alternate Forum for Research in Mindanao.

Gomes, Robin. 2021. "Philippine Church Decries Government Lifting Ban on Mining." *Vatican News*, April 21, 2021. https://www.vaticannews.va/en/church/news/2021-04/philippines-caritas-bishops-against-lifting-ban-mining.html.

Goodland, Robert, and Clive Wicks. 2008. *Philippines: Mining or Food?: Chapters 1 to 7*. London: The Working Group on Mining in the Philippines. http://www.piplinks.org/miningorfood.html.

Holden, William N., and R. Daniel Jacobson. 2007. "Mining Amid Armed Conflict: Nonferrous Metals Mining in the Philippines." *The Canadian Geographer* 51, no. 4: 475–500. 10.1111/j.1541-0064.2007.00193.x.

Odchimar, Nerio. 2010. Nerio Odchimar, Bishop of Tandag, to Benigno Simeon C. Aquino, III, President of the Republic of Philippines. July 12, 2010.

Reid, Ben. 2000. *Philippine Left: Political Crisis and Social Change*. Manila: Journal of Contemporary Asia Publishers.

Sanz, Penelope (coordinator). 2001. "Philippines: Mining a Sacred Mountain: Protecting the Human Rights of Indigenous Peoples." Human Rights Impact Assessments for Foreign Investment Projects. https://www.slideshare.net/no2mininginpalawan/mining-a-sacred-mountainprotecting-the-human-rights-of-indigenous-communities.

Shepherd, Bede, and A. Lin Neumann. 2007. "Scared Silent: Impunity for Extrajudicial Killings." *Human Rights Watch* 19, no. 9. https://www.hrw.org/reports/2007/philippines0607.

Sison, Jose Maria. 2013. *Detention and Defiance against Dictatorship: Selected Writings, 1977–1986*. Utrecht, The Netherlands: International Network for Philippine Studies.

Vitug, Marites Danguilan, and Glenda M. Gloria. 2000. *Under the Crescent Moon: Rebellion in Mindanao*. Quezon City: Ateneo Center for Social Policy and Public Affairs.

6 Dynamics between the state, mining companies, and indigenous peoples in Peru

José Bayardo Chata Pacoricona[1]

The business sector is adamant: mining is an essential activity for the future of Peru. The impact of mining on sectors like metalworking, agriculture, livestock, fishing, information technology, and communications purportedly demonstrates the importance of this sector for the country's growth (see Benavides Ganoza 2018). However, at the community level, these positive impacts are rarely felt, and frequently communities' dissatisfaction about being shut out from the development of a mining project leads to social protests and other forms of resistance that are ripe for violence and conflict.

A good example is Tía María, an open-pit copper mine in the Cocachacra district of Peru's Islay Province that has led to a decade of conflict with the local community. The government has strongly supported this project despite strong opposition from farmers, whose access to groundwater is threatened by the preferential water access given to the mining company. In response to social conflict, the company eventually reversed course and decided to use seawater, but the water supply remains at risk of contamination. Despite this, the government has persisted in its position that the project is vital to national economic growth.

The main actors in all socio-environmental conflicts around mining, like Tía María, are the same. First, we find the people, whose rights are overlooked as they are forced into a damaging economic activity alien to the cultures and livelihoods to which they are accustomed. Secondly, there is the private company, which "in compliance" with all the requirements imposed by national legislation for undertaking extractive activities, struggles to make effective its claim to extract minerals from the subsoil. And finally, there is the state, with its pro-extractive and pro-private investment policies, regulations, and practices that do not align with its obligations to defend its citizens, guarantee human rights, and promote general well-being founded on justice (see Political Constitution of Peru, art. 44).

Our organization, Derechos Humanos y Medio Ambiente—Puno, Peru (DHUMA), pressures the government to meet its obligations by pushing back against deceptive narratives about positive impacts of mining and

DOI: 10.4324/9781003094272-6

helping communities resist mining development through legal processes that they are typically unaware of or not capable of utilizing. DHUMA was founded in 1988 as the Vicariate of Solidarity within the Catholic Prelature of Juli. The impetus was the increasing political violence faced by indigenous communities during the 1980s. Several religious congregations, lay parish workers, and ecclesial leadership coordinated to found the Vicariate, and in 2008 it became an independent civil society organization. DHUMA has maintained its Catholic identity and its mission remains rooted in the principles of the gospel and Catholic social teaching. And its work has remained focused on helping the indigenous communities of the Puno region.

This chapter describes the fraught dynamics between communities, companies, and the state in the context of mining, and demonstrates how we attempt to defend those communities before these situations lead to violence, dispossession, ecological destruction, and human rights abuses.

The state and the corporations

The bonds between private companies and the state are extremely worrying. The dealings between private investment and the government are principally based on lobbying agendas, hidden negotiations, and corruption. For example, the Lava Jato corruption case is one of the largest known to date, involving private companies and governments in various countries across Latin America where infrastructure megaprojects have been carried out on the back of billions of dollars in bribes. According to a report from Oxfam:

> The analysis of these megaprojects allows us to understand the way in which laws are made, influencing what is called 'tailored legislation,' legislation tailored to certain interests in how it is created and applied, where governments and officials frequently ignore, especially in cases of public works, the spirit of the laws and their necessary ethical basis, attending instead to the mere fulfillment of formalities to give a legal varnish to projects that have numerous problems. (Durand 2018, 17)

This appearance of legality and claims that the state is fulfilling its duty to promote general well-being masks the real goal behind each project or private investment: the self-interest of politicians and public officials.

Ordinary citizens are bystanders to these million-dollar scandals and are left wondering if the whole state machinery moves only to the rhythm of monetary kickbacks and whether all the laws and policies of the state work like this. In the same way that there is empirical certainty about institutionalized corruption in the Peruvian judicial system (IDL Reporteros 2018), illegal agreements made between private companies and the Peruvian state are being revealed as they unravel over time.

An emblematic case: the Aymarazo

On November 28, 2007, Supreme Decree 083–2007-EM was issued, declaring private investment in mining a public necessity. It enabled the Peruvian subsidiariy of Canada's Bear Creek Mining Corporation to acquire concessions and rights over mines and complementary resources for development of production activities inside a fifty-kilometer zone along the border with Bolivia. This development, the Santa Ana mining project, is located in the Puno region on indigenous Aymara territory. This decree was the starting point for the socio-environmental conflict known as the Aymarazo.

To deliver the concessions that make up the Santa Ana project, an exception had to be used. Constitutionally, foreigners cannot acquire land rights of any kind within fifty kilometers of the border (Political Constitution of Peru, art. 71). Upon signing the supreme decree, this article of the constitution was dismissed on the grounds of "public necessity." In issuing this decree, the state sought to substantiate that granting mining rights to Bear Creek would, in fact, have a beneficial impact on the well-being of the general community. In addition, it was assumed that development of this border zone would bring improved living standards for the inhabitants in the area—something that, unfortunately, generally does not happen in these situations. "Public necessity" should be about needed works that are truly public, that benefit the broader community and not just private business interests. The "public necessity" designation is, for example, for infrastructure works like the construction of roads, hospitals, or dams; the decontamination of lakes or rivers; or the extension of electricity, telecommunication, or drinking water networks (Ruiz Molleda 2011). As such, we question what this business activity has to do with the public interest.

Agreements between the National Police of Peru and mining companies

In 2019, three institutions looked closely into the agreements that have been signed between the National Police of Peru and mining companies nationwide. The examination covered 138 agreements signed between 1995 and 2018. In the report, these institutions discovered that 109 of the agreements were signed before 2017, an average of nine per year. Then the count accelerated to 29 between 2017 and 2018. On review, it seems that 29 agreements in force at the time were due to expire between 2019 and 2022. What is striking is that most of the existing agreements apply to projects located in regions where there has been social unrest and the projects have been questioned by local communities (EarthRights International, Instituto de Defensa Legal, and Coordinadora Nacional de Derechos Humanos International 2019).

The content of these agreements deserves special study since they differ from each other in their commitments. There are agreements whose aims include the "detection and neutralization of risks," without establishing any exact definition, leaving the determination open to interpretation by the contracted police officers and the corporate representatives. "Risk" could be read as any action, such as a demonstration or protest, carried out by a person or group defending interests or opinions that differ from those of the corporations concerned. Other agreements have clauses guaranteeing "normal development," and controlling "acts of vandalism and sabotage or terrorism." These ambiguous and indeterminate terms can be interpreted in a discretionary way by interested parties, favoring private interests to the detriment of the individual and collective rights of the local population (EarthRights International, Instituto de Defensa Legal, and Coordinadora Nacional de Derechos Humanos International 2019).

As a result of these agreements between the National Police of Peru and extractive corporations, the police have been providing extra services biased toward corporate interests and discriminating against communities in exchange for economic compensation. In this way, police officers have become private guards for corporations, undermining their function as public servants who receive their salaries from taxpayers. In practice, we are faced with the privatization of domestic order and the public function of the police force (Ruiz Molleda and Salvador n.d.).

This situation makes clear an obvious conflict of interest for the National Police. It is impossible for its forces to distinguish public from corporate interest. This scenario encourages criminalization of social protest and abusive use of police force. Ultimately, the damage is borne by indigenous communities and peoples that live in the immediate vicinity of mining operations (Leyva 2018, 39). The crisis worsens in situations of social conflict, where the partiality of the National Police can result, and has resulted, in clear human rights violations. This is demonstrated by the case of the Las Bambas copper mine in Apurímac, where, since 2015, criminalization of the rural population has resulted in four protestors' deaths. This police "function" is obviously not protecting public order but rather the interests of corporations; it is not preserving peace, but is aggravating conflict (EarthRights International, Instituto de Defensa Legal, and Coordinadora Nacional de Derechos Humanos International 2019).

The state and indigenous peoples

The idea of development from two perspectives

The Aymarazo case is an example of a clash of perspectives on development. By declaring the mine a public necessity, and in granting mining concessions that skirted constitutional standards, the state asserted that the mining project would contribute to economic development and to the improvement of the quality of life of the rural communities involved. The conviction of the state remains that mining should be at the center of the national economy.[2] As such, it places the entire state apparatus at the service of its mining portfolio, without considering the right of communities or ancestral owners to refuse mining in their lands and territories. Whether or not the government is acting in good faith, the reality is that this vision of progress differs greatly from the ancestral perspective of development that communities hold. That view is rooted in a sacred understanding that cannot be adequately captured in economic or patrimonial reckonings alone, and that resonates with Catholic senses of care for creation and integral ecology (Francis 2015, ch. 4). The result of these opposing positions is socio-environmental conflict. This pattern is repeated in the majority of such conflicts in the country. Despite the omnipresence of these dynamics, it appears there is no real solution proposed that could meaningfully prevent future conflicts. In June 2019, the Ombudsman's Office of Peru, Office for the Prevention of Social Conflicts and Governance (2019) reported four new social conflicts in the prior month, raising the national total to 178: 130 active and 48 latent. Of the 178, 117 were of a socio-environmental nature, 62.4% (73 cases) of which were related to mining activity and 16.2% (19 cases) linked to hydrocarbon activity. This data shows the level of resistance to mining and hydrocarbon activities as expressed through social protest.

This final point has to do with the right of the people to self-determination. With its pro-extractivist vision, the state arbitrarily wields the power to decide what is beneficial for indigenous peoples (IPs), completely distorting this right. Article 7 of Convention 169 of the International Labour Organization indicates that IPs have the right to decide their own priorities regarding economic development. In Catholic social teaching, the right of self-determination is grounded particularly in freedom and the right of individuals and peoples to participate in shaping their futures. Straightforwardly, Pope Francis (2020, §14) has written that business enterprises that violate the self-determination of IPs are crimes. Respect for self-determination is necessary for integral human development, defined not only in economic terms, but also in terms of protecting communal identity and enabling all peoples to contribute to the common good (Powers 2019, 60).

In 2019, Peru's Ministry of Energy and Mines released "Vision of Mining in Peru to 2030," outlining a path for making mining a central pillar of the national economy.[3] The dependence of our economy on mining is worrying, and is reinforced through norms and state policies that reaffirm the extractive-oriented interests and priorities of corporations and the government. The rapid change in national regulations and state policies is clearly at the expense of the people who already live in mining areas, and IPs who are at risk of having their lands lost to mining. These changes seek to eliminate, in a way that is presented as beneficial, the guarantees of communities' rights to land and territory, an adequate and balanced environment, and participation, prior consultation, and fundamental self-determination about their development. All of this goes against the nation's constitution, international human rights norms, and principles of Catholic social thought and integral human development.

Policies and governance favoring mining companies over communities

Law 30230 places more than eight thousand rural and indigenous communities in a situation of vulnerability. This law allows the state, through "special procedures," to hand over land to investment projects, in the location and amount that the investor requires, ignoring the property rights of legitimate owners. For mining, these owners are often IPs. These special procedures affect all properties, be they private or state lands, whether they are backed by deeds or any other documentation. No exceptions of any kind apply.

One of the legislative decrees most responsible for weakening communal property rights and the autonomy of rural communities has been D.Leg. 1333, which is known as the "law of dispossession." This law, which builds upon Law 30230, was repealed in Congress, but it is kept in force by executive action. The regulation allows the state more control over communal property and territory so as to facilitate granting mining concessions to private companies. This standard is complemented by D.Leg. 1320, which extends the ownership of mining concessions to thirty years, even for concession holders who do not pay their annual right of validity fee to the state (Red Muqui n.d.).

Laws and policies prioritizing mining by weakening environmental regulation have also been harmful to IPs. The loosening of environmental protection standards is based on modifications to the government's environmental quality standards for air and soil, and the maximum permissible limits of toxic elements. These modifications give wide latitude for companies to pollute while offering virtual impunity and preventing compensation for those communities harmed by pollution. Among these rules is Supreme Decree 003–2017. This decree weakens standards for sulfur dioxide and other toxic materials generated by extraction activities. Chief Resolution 056–2018-ANA on the Classification of Continental Water Bodies is another example. It was meant to designate some bodies of water as more fragile and thus due stricter protection. But technical clarification on the classification was not given, allowing mining companies to more easily pollute. Other regulations have been issued that favor the privatization of water (D.Leg. 1280), facilitating its privileged use for mining activity and not for citizens' collectives or consumption by the general population. Legislative Decree 1285, issued at the end of 2016, modifies article 79 of Law No. 29338, the Law of Water Resources. These norms establish provisions for a progressive permissiveness towards oil spills and mining tailings and make environmental management mechanisms more malleable, during and after extractive activity. In short, these provisions facilitate pollution with impunity (Red Muqui n.d.).

Another tactic the government has used to harm IPs and other communities is declaring states of emergency. The state has used this tactic to restrict fundamental rights and allow police or military intervention without having to prove the existence of a serious breach of public order, as required by article 137 of the Peruvian constitution. This is a blatant abuse of state power outside the constitutional framework. For example, a state of emergency was imposed in the districts tied to the Las Bambas mining project that included the entire Apurimac-Cusco-Arequipa Road Corridor, which is 482 square kilometers in area and extends 500 meters on each side of the highway. In some places, states of emergency have been declared preventively, restricting such fundamental rights as individual freedom, freedom of assembly and movement, and inviolability of the home. In some cases, citizens have been prevented from having public gatherings, even ones that are not of a political nature (Ruiz Molleda et al. 2019).

Consultation, social protest, and peaceful remediation

Formal processes have always presented an obstacle for indigenous peoples. Mining concessions are unilaterally granted by the Mining and Metallurgical Geological Institute (INGEMMET) at the request of an interested corporation.[4] Consequently, indigenous communities based at mineral-rich sites live in constant fear, under the permanent threat that hangs over their land. According to law, the mining concession holder will not be able to use the land unless he or she has an official agreement with the land owner or has been granted a mining easement. However, if no agreement has been reached, the holder of the mining concession may request an easement from the General Mining Directorate of the Ministry of Energy and Mines. In other words, even in the face of opposition from the community, the legal framework favors the concession holder. A concession holder can press ahead with mining exploration and exploitation by seeking an easement, regardless of whether the community opposes the project.

Convention 169 of the International Labour Organization (1989), which grants protections to IPs, was ratified by Peru in 1994. The struggle to uphold Convention 169 is led by the same people it is meant to protect, with accompaniment from us at DHUMA, but without support from the state. Although progress is being made in short steps for the rights contained in this rule to be recognized in practice, the state continues by various means to undermine efforts to enforce it. For example, the peasant communities of Chila Chambilla and Chila Pucara were the first Aymara indigenous communities to initiate a judicial process against the Peruvian state for the omission of the right to prior consultation and the violation of the right to territory in the delivery of mining concessions to a private company. The magistrates in charge of the case stated that Convention 169 does not protect them because they were unable to provide sufficient proof that they belong to an indigenous people, denying them their cultural identity.

Article 6.1.a of Convention 169 makes clear that consultation with IPs about use of their land needs to happen in the earliest stages of any possible project: "governments should ... consult the peoples concerned, through appropriate procedures and in particular through their representative institutions, whenever legislative or administrative measures are expected to directly affect them." This standard has been developed by article 4.a of the Law on Consultation (Law 29785), and a similar concept, called the Constitutional Principle of Prior Implementation of the Consultation Process, has been proposed by Peru's Constitutional Court. This principle mandates that "consultation be carried out prior to any decision. An essential idea behind the inclusion of indigenous peoples in the discussion of the project of the administrative or legislative measure is that they have an opportunity to raise their cultural perspectives, so that they may be taken into account" (Constitutional Tribunal of Peru 2010, legal basis 36).

But the procedure for granting mining concessions in Peru is absolutely incompatible with these legal norms (see International Labour Organization 1989, art. 13–15; Political Constitution of Peru, art. 88). The Aymarazo conflict arose when rural communities in Puno learned that INGEMMET had delivered mining concessions on about 60% of the territory of Puno, behind the backs of communities and without their prior consultation. This follows a typical pattern where communities often become aware of concessions only after the deadline to file their opposition has passed. Prior consultation can be used to channel misgivings and avoid social protests by communities, but it is frequently evaded or enacted in only superficial ways.

The one requirement to be granted a mining concession is that the potential concession holder publish a single notice in two newspapers, one of national and the other of regional scope. Five key barriers prevent rural communities from being able to access such notices: (1) notices are published in newspapers that only circulate in urban areas and not in rural areas where IPs live; (2) they are published in Spanish and not the native languages of IPs; (3) the only way to access these notices is to buy the newspapers, which are expensive for the economically vulnerable IPs; (4) Andean cultures are mostly oral, which prevents indigenous communities from understanding the content of the ads; and (5) complete information relating to the mining concessions and their respective administrative files are logged in a database which can only be accessed through a computer with an internet connection, which many indigenous communities do not have.[5] And in the event that a notice does reach the communities, it does not specify the name of the communities to be affected, or include maps of the concessions. This all means that IPs are being systematically excluded from legal processes for determining the status of their lands. However, the General Law of Administrative Procedures is very clear about what ought to happen: "If, during the filing of a procedure, the existence comes to light of non-appearing third parties whose legitimate rights or interests may be affected by the resolution to be issued, said procedure must be communicated to them by means of a summons delivered to their known address, without interrupting proceedings" (Presidency of the Council of Ministers of Peru 2001, art. 60.1). This legal provision has not been complied with in any mining administrative procedure.

Faced with this lack of meaningful prior consultation, and the frequent failure of the state to take into account the livelihoods and cultures of native peoples, archaeological remains,[6] or areas destined for agriculture expansion[7] when granting concessions for mining, communities choose to exercise their right to social protest. This leads to conflict and opens the way for violence, especially in light of the way social protest has been criminalized.

The criminalization of protest is a multidimensional phenomenon that includes actions and speech aimed at repressing and delegitimizing political dissent. Acts of repression may include assassinations, executions, disappearances, attacks, threats, harassment, spying, and persecution via criminal proceedings, and they may be against an individual or a collective. Criminalizing discourse discredits protesters as lawbreakers, radicals, and, in the most extreme cases, terrorists. This is the ideological justification that favors the state against social protest (Saldaña and Portocarrero 2017).

Social protest is distorted by the state and treated as a criminal offense to be punished. A clear example of the use of law to prosecute protest leaders is article 200 of the Peruvian Criminal Code, third paragraph:

> The one who, through violence or threat, occupies premises, hinders communication channels or prevents the free transit of citizens or disrupts the normal functioning of public services or the execution of legally authorized works in order to obtain any benefit or advantage from the authorities of an undue economic or other advantage of any other nature ... The penalty shall be not less than fifteen or greater than twenty-five years [of imprisonment] and disqualification pursuant to numerals 4 and 6 of article 36, if the violence or threat is committed. (Ministry of Justice and Human Rights of Peru 2016)

Following this logic, the main Aymara spokespeople saw themselves criminally denounced for crimes such as riots, aggravated extortion, and obstruction of public services following the Aymarazo protest. Such criminal definitions place a chilling pall over any act of protest. In joint inter-institutional cooperation, the Institute of Legal Defense and the Bar

Association of Puno have sought a declaration of unconstitutionality of this article so that it ceases to have criminalizing effects.

In light of these realities, our organization attempts to offer resources for communities to use legal mechanisms as a means of peacefully attaining justice and realizing their own visions of development. We see this alternative as urgent, recognizing that protest can result in injury, death, detention, and criminal charges. In that same vein, human rights organizations decided to present a set of lawsuits before the judiciary and the Constitutional Court against different state actors for their failure to comply with consultation requirements. Such claims form part of what we call "strategic litigation." On top of obtaining favorable sentences for individual cases, we seek to put discussions regarding the situation of rural communities in extractive contexts high on the legislative agenda. Cases where consultation was omitted for mining activity in Puno include the Quechua communities of Atuncolla, San José de Llungo, San José Principio, and Arboleda, which was a pioneer in demanding consultation at the national level, and the Aymara communities of Jatucachi, Chila Chambilla, and Chila Pucara.

Conclusion

When we examine the dynamics between the state, companies, and indigenous communities, it appears that IPs have no alliances other than some select groups dedicated to advocating for them. Through corruption, legal and procedural complexity and obfuscation, and criminalization of protest, IPs cease to be protagonists in the stories of their own lands. They are beset by grave power imbalances and other disadvantages that prevent them from being able to use legal channels before mining projects move past critical points of development. And when they try to respond after the fact, they are frequently met by violence and repression.

The social doctrine of the Catholic Church teaches us to take a political, social, and religious stance in favor of the people, especially the most vulnerable. This doctrine encompasses all aspects of human dignity, and its fulfillment is central to the Christian vocation. At DHUMA, we focus on the mandate in *Laudato Si'* to hear the cry of the earth as well as the cry of the poor. From our time as part of the Prelature of Juli through our work today as an independent organization, we continue to dedicate ourselves to the defense of life and Mother Earth, and the accompaniment of the children of God. In Puno, that mostly includes dealing with the violation of the rights of IPs in relation to mining. The long history of suffering of native peoples was aggravated greatly during the political violence in Peru from 1980 to 2000, and now that suffering includes violence against nature and lands which IPs hold sacred and on which they depend for their well-being.

Because of this reality and our commitment to Catholic social teaching, we help indigenous communities know their rights, understand mining development projects and processes, and navigate legal channels to redress problems. Such activities are an important element of peacebuilding in Peru and other countries in the Global South facing similar challenges, whether they involve IPs or other vulnerable and marginalized communities. DHUMA, rooted in a Catholic legacy, will continue to advance and deepen such work with the communities of Puno, Peru.

Notes

1 Translated from Spanish by Thomas McDonagh.
2 The Directorate General for Promotion of Sustainable Mining presented a 2020 update of the Mine Construction Project Portfolio consisting of 46 projects and investments amounting to US$56.158 million. These investments include projects for the construction of new mines, expansion of existing ones, and the reuse of tailings (Ministry of Energy and Mines of Peru 2020).
3 DHUMA is a member of Red Muqui, a national platform of institutions that engages in advocacy to support rural communities and defend the environment in the context of mining. Red Muqui refused to endorse this statement, despite its attempts to present sustainable and inclusive models of mining development.
4 INGEMMET is a decentralized technical public body that forms part of the Ministry of Energy and Mines. INGEMMET's work is aimed at obtaining, administering, and efficiently disseminating geoscientific information and data related to basic geology, subsoil resources, and geological risks. It is also responsible for conducting the Ordinary Mining Procedure, including receiving petitions, granting and revoking mining concessions, systematizing georeferenced information through the National Mining Cadastre, as well as the administration and distribution of permits and penalties.
5 Despite this, curiously, INGEMMET has been awarded prizes by the state. In 2012, it achieved first place in its category, Transparency and Access to Information, based on the use of information systems called GEOCATMIN (Geological Cadastral Mining System) and SIDEMCAT (Mining Rights and Cadastre System) and on its role as a disseminator of the geological and cadastral information that the institution manages. Also, at the 2015 conference "Good Practices in Geo-Information Management" organized by the National Office of Electronic Government and Information Technology, INGEMMET won an award in the category of Metadata Management.
6 In 2011, INGEMMET delivered mining concessions in grids that covered the pre-Inca archaeological complex of Sillustani, whose main attraction are its chullpas, funeral towers that formed part of the architectural expression of the Kolla culture, located in the Atuncolla district, Puno province and region.
7 Article 14 of the General Mining Law indicates that: "... no non-metallic concessions or extensions of non-metallic concessions may be established, on intangible agricultural areas, or on rustic lands for agricultural use, without considering between these and last to natural pastures" (Ministry of Energy and Mines of Peru 1992). However, the Ministry of Agriculture and Irrigation does not have official data on all the territories designated for agriculture, and many communities have extended agricultural activity beyond known territories. And in the case of mining concessions of a metallic nature, no limit is established, whether it is for agricultural purposes or not.

References

Benavides Ganoza, Roque. 2018. *La Minería Responsable y sus Aportes al Desarrollo del Perú*. 3rd ed. Compañía de Minas Buenaventura. https://www.buenaventura.com/assets/mineria-responsable/#p=1.

Constitutional Tribunal of Peru. 2010. Sentence of the Constitutional Tribunal, Exp. N° 00022–2009-PI/TC. https://www.tc.gob.pe/jurisprudencia/2010/00022–2009-AI.html.

Durand, Francisco. 2018. *Odebrecht—La empresa que captura gobiernos*. Lima: Pontificia Universidad Católica del Perú; Oxfam Peru. https://oi-files-cng-prod.s3.amazonaws.com/peru.oxfam.org/s3fs-public/file_attachments/Odebrecht-La-empresa-que-capturaba-gobiernos.pdf.

EarthRights International, Instituto de Defensa Legal, and Coordinadora Nacional de Derechos Humanos International. 2019. *Informe: Convenios entre la Policía Nacional y las empresas extractivas en el Perú: Análisis de las relaciones que permiten la violación de los derechos humanos y quiebran los principios del Estado democrático de Derecho*. https://www.idl.org.pe/portfolio/convenios-entre-la-policia-nacional-y-las-empresas-extractivas-en-el-peru.

Francis. 2015. *Laudato Si'*. http://www.vatican.va/content/francesco/en/encyclicals/documents/papa-francesco_20150524_enciclica-laudato-si.html.

Francis. 2020. *Querida Amazonia*. http://www.vatican.va/content/francesco/en/apost_exhortations/documents/papa-francesco_esortazione-ap_20200202_querida-amazonia.html.

IDL Reporteros. 2018. "Corte y corrupcíon." Updated July 12, 2018. https://www.idl-reporteros.pe/corte-y-corrupcion/.

International Labour Organization. 1989. *Indigenous and Tribal Peoples' Convention*. C169. https://www.ilo.org/dyn/normlex/en/f?p=NORMLEXPUB:12100:0::NO::P12100_ILO_CODE:C169.

Leyva, Ana. 2018. *Consúltame de verdad. Aproximación a un balance sobre consulta previa en el Perú en los sectores minero e hidrocarburífero*. Lima: Oxfam, CooperAcción. http://cooperaccion.org.pe/wp-content/uploads/2018/07/Consultame-de-verdad.pdf.

Ministry of Energy and Mines of Peru. 1992. General Mining Law (Supreme Decree N° 014–92-EM). http://www.minem.gob.pe/minem/archivos/file/Mineria/LEGISLACION/TUO%20.pdf.

Ministry of Energy and Mines of Peru. 2019. "Visión de la Minería en el Perú al 2030." http://www.minem.gob.pe/_detalle.php?idSector=1&idTitular=9757&idMenu=sub149&idCateg=1884.

Ministry of Energy and Mines of Peru. 2020. *Cartera De Proyectos De Construcción De Mina*. http://www.minem.gob.pe/minem/archivos/file/Mineria/INVERSION/2020/CPM2020(1).pdf.

Ministry of Justice and Human Rights of Peru. 2016. Penal Code (Legislative Decree N° 635). http://spij.minjus.gob.pe/content/publicaciones_oficiales/img/CODIGOPENAL.pdf.

Ombudsman's Office of Peru, Office for the Prevention of Social Conflicts and Governance. 2019. *Reporte de Conflictos Sociales N° 184*. https://www.defensoria.gob.pe/wp-content/uploads/2019/07/Conflictos-Sociales-N%C2%B0–184-Junio-2019.pdf.

Political Constitution of Peru. http://www.congreso.gob.pe/Docs/files/constitucion/constitucion2019/index.html.
Powers, Gerard. 2019. "Toward a Catholic Ethic of Self-Determination." In *Theology, Conflict, and Peacebuilding*, edited by Daniel Franklin Pilario, 39–74. Quezon City, Philippines: St. Vincent School of Theology. https://www.svst.edu.ph/publication-review/603f3c455c60e245#.
Presidency of the Council of Ministers of Peru. 2001. Law of General Administrative Procedures. Law N° 27444. http://www.pcm.gob.pe/wp-content/uploads/2013/09/Ley-de-Procedimiento-Administrativo-de-PersonalLey27444.pdf.
Red Muqui. n.d. "Normas y políticas públicas que afectan derechos fundamentales, colectivos y ambientales 2016–2018." Accessed April 13, 2021. https://muqui.org/wp-content/uploads/2019/11/Normas-y-politicas-publicas-2016–2018-IMAGENES-2.pdf.
Ruiz Molleda, Juan Carlos. 2011. "Análisis jurídico de las normas legales expedidas por el gobierno en el marco de las negociaciones con las comunidades campesinas de Puno." Institut de Defensa Legal —Justicia Viva, Working Paper No. 53.
Ruiz Molleda, Juan Carlos, and Álvaro Másquez Salvador. n.d. "Informe juridico: la privatización de la seguridad ciudadana: La inconstitucionalidad de los contratos de seguridad privada entre empresas mineras y la Policía Nacional del Perú." Justicia Viva. Accessed April13, 2021. http://www.justiciaviva.org.pe/new/wp-content/uploads/2018/05/Informe-Jur%C3%ADdico-sobre-la-privatizaci%C3%B3n-del-orden-interno.-Versi%C3%B3n-final-16-09-2015.pdf.
Ruiz Molleda, Juan Carlos, Rocío Meza Suarez, Marisa Quispe Mamani, Álvaro Másquez Salvador, and Isaac Peña Lobato. 2019. "¿En qué va el litigio constitucional contra la criminalización de la protesta?" Instituto de Defensa Legal. June 18, 2019. https://idl.org.pe/en-que-va-el-litigio-constitucional-contra-la-criminalizacion-de-la-protesta/.
Saldaña, José, and Jorge Portocarrero. 2017. "La violencia de las leyes: el uso de la fuerza y la criminalización de protestas socioambientales en el Perú." *Revista de la Facultad de derechos de la Pontificia universidad Católica del Perú* 79: 311–352. http://revistas.pucp.edu.pe/index.php/derechopucp/article/view/19329/19456.

7 The mining struggle in El Salvador and the role of the Catholic Church

Andrés McKinley

In the middle of the sixteenth century, when King Ferdinand of Spain was orienting his men on the priorities for the conquest of the new world, his instructions were clear and concise: "Get the gold, humanely if possible; but, at any cost, get the gold" (Perlez and Johnson 2005, n.p.). Five hundred years later, following in the footsteps of the conquistadores, transnational mining corporations from Canada, the United States, Australia, and as far away as Russia, are coming to Central America in search of gold, silver, iron, nickel, and other valuable metals in a mineral belt extending from Panama to Mexico.

In an effort to win the hearts and minds of governments and local communities, these corporations promise new technologies which, they claim, protect the environment, utilizing terms like "modern mining" or "responsible mining." They promise jobs and an economic boom for poor and desperate communities located near mineral deposits. They assure a significant increase in government income which can then be used to support health, education, and other national priorities, and promise respect for basic human rights. In spite of this enormously attractive offer, the people of El Salvador said "no."

Public opinion polls carried out by the Jesuit-run Central American University (UCA) in 2007 and 2015 demonstrated a massive rejection of metallic mining by affected communities and broad sectors of Salvadoran society. In 2015, 79.5% of those polled considered El Salvador to be an inappropriate place for metallic mining. Over 76% expressed opposition to potential mining projects in their municipality and 77% demanded that the government take immediate measures to prohibit metallic mining. Through the work of UCA and informed and steadfast support from the country's episcopal leadership, the Catholic Church was one of the leading forces for translating that widespread concern about mining into national policy for the common good of the people. In what follows, I will describe the main reason why El Salvador's people rejected mining: water. I will then outline the country's history of resistance to mining and analyze why it ultimately proved successful. Finally, I will address the distinct and central role played by the Catholic Church in this struggle.

DOI: 10.4324/9781003094272-7

Why El Salvador said "no" to metallic mining

How can one understand the radical position against mining taken by the citizens and the Catholic Church of a small impoverished country in dire need of direct foreign investment, jobs, and better services in health and education? The answer to this question can be summarized in a single word: water.

El Salvador suffers from a severe crisis in terms of quantity, quality, and access to water. The crisis has its origins in decades of mismanagement, overexploitation, and contamination. Recent studies indicate that if the country does not make profound changes in how it manages water resources, life in El Salvador will be unviable within eighty years (Agencia EFE 2016). And the situation continues to worsen. In April of 2019, the Ministry of Environment and Natural Resources (MARN) gave a dismal account of the prospects for freshwater availability over the first three months of the rainy season (May–July). This prediction was based on the findings of the recent Central America Climate Forum in which rain patterns were analyzed and compared with previous years. In its report, MARN revealed that the average water flows of several of the most strategic rivers in the country would show a deficit of between 45% and 60% between the months of May and July. The Huiza, Chilama, and Amayo rivers, in the province of La Libertad in central El Salvador, would reduce their water flows by up to 50%. In the eastern part of the country, the Goascorán river in the province of La Unión, the Torola river in the department of Morazán, the Rio Grande in San Miguel, and the Molino and Cacao Rivers of Usulután would be reduced by more than 60%. Finally, in the western zone, the flow levels of the Angue, San José, and Ostúa Rivers of Metapán in the province of Santa Ana would see reductions of 45%.

According to MARN, aquifers around the country are also being gradually depleted by climate change and overexploitation by large-scale agriculture and industry. The impact has been dramatic, with rural communities suffering from decreases in water availability in natural springs and wells, and with aquifers that supply potable water to important urban centers reaching worrisome levels. Aquifers supplying water to the capital city of San Salvador suffered a decrease of 4.23 meters between 2012 and 2019. The aquifers supplying water to the eastern city of San Miguel suffered a similar decrease (4.14 meters) between 2014 and 2019, and the aquifer of San Agustín-San Francisco Javier in the southern city of Usulután recorded a decrease of 4.44 meters between 2013 and 2019 (Calderón 2019). According to the government's National Administration of Aqueducts and Sewers, no less than 45 wells providing potable drinking water for 2.1 million people in greater San Salvador were depleted between 2008 and 2016, and water levels in remaining wells are diminishing at a rate of between

1 and 1.5 meters per year (Machuca 2019). Adding to the crisis, environmentalists have reported that over 90% of surface waters suffer from some degree of contamination (La Prensa Gráfica 2016).

These developments have led highly respected international experts like the UN Economic Commission for Latin America and the Caribbean, the Latin American Water Tribunal, and the Global Water Partnership to declare that El Salvador is on the threshold of water stress, a situation in which freshwater resources available for human use decline to a level of 1,700 cubic meters per person per year, rendering it impossible to meet demand. MARN, for its part, warns that by 2022, 80% of El Salvador´s national territory will suffer from water stress (Agencia EFE 2019).

Metallic mining requires enormous quantities of freshwater. One can say that freshwater is the lifeblood of mining in the same way that it is the lifeblood of human existence. Mining competes with local communities, and with humanity at large, for this vital liquid. The average gold mine in Central America utilizes more than a million liters of water per day. Some mines in the region utilize more than 6 million liters per day, and larger mines around the world utilize hundreds of millions (McKinley 2016, i, 21). Metallic mining, especially for gold, also contaminates water. The separation of gold from ore requires sodium cyanide, a toxic chemical that can be lethal in quantities smaller than a grain of rice. Cyanide solutions frequently seep into freshwater sources in and around mining sites, or escape from poorly constructed tailings dams where toxic waste from the mining process is stored (24–25). Cyanide also evaporates at 36°C, contaminating the air in a radius of many miles around mining operations.

Apart from sodium cyanide, metallic mining utilizes explosives, fuel, antifreeze, and other materials that leave toxic residues that seep into freshwater systems. The most problematic source of water pollution from metallic mining, however, is a process called acid mine drainage, a phenomenon that occurs when rock with high sulfide content (commonly found in Central America) is extracted from the earth, crushed, and exposed to oxygen from air and rainwater, which converts the sulfides into sulfates and, finally, into sulfuric acid. The sulfuric acid, as well as other harmful metals that it leaches from the walls of the mine and rocks in the surrounding area, flows into nearby streams, aquifers, and other freshwater sources. Acid mine drainage is a difficult process to reverse and can be found today in gold mines in France and Spain that date to the time of the Roman Empire (Rodriguez, Garcia, and Castillo 2007, 33).

Given these factors, it is obvious that a country like El Salvador—on the brink of water stress, with high levels of pollution, the highest population density in the western hemisphere, and already suffering from severe water shortages—is not an appropriate location for large-scale metallic mining.

The national struggle against metallic mining

The struggle against metallic mining in El Salvador lasted for over seventeen years and evolved through four clear and distinct phases.

Phase 1 (2000–2004) was marked by the incursion of transnational mining companies and the growing resistance of local communities. This phenomenon was especially clear in the northern provinces of Chalatenango and Cabañas, regions highly impacted by the civil war of 1980–1992. In the case of Chalatenango, the communities most threatened by mining were well-organized and committed to protecting their territories, considered sacred and drenched in the blood of family members assassinated by the army during the war. These communities were quick to respond to the presence of foreign mining companies prospecting for gold and silver, and they built strong local processes of resistance.

The case of Cabañas was more complex. The communities in this province had been polarized by the war, with ex-combatants from guerrilla organizations living side by side with former soldiers or collaborators with the army. They were engaged in a painful search to heal the wounds of twelve years of war and mend the social fabric of their communities. Different opinions about mining exacerbated existing divisions and hindered local resistance. Nevertheless, the province, like Chalatenango, remained predominantly anti-mining.

Phase 2 (2005–2009) was characterized principally by the formation and consolidation of a national struggle against metallic mining based on the logic of public policy advocacy. This initiative began in 2005 with the organization of a broad coalition, the National Roundtable on Metallic Mining in El Salvador (MESA), consisting of community-based organizations, development NGOs, environmentalists, church groups, legal aid organizations, and other actors. MESA, strongly supported by Oxfam America and other international organizations, quickly became the backbone of the struggle. It employed a rights-based focus, promoted concrete policy proposals, and designed and applied well-planned advocacy campaigns to advance its policy agenda.

Phase 3 (2009–2016) was dominated principally by a prolonged legal struggle which began with a promise by President Antonio Saca in 2009 that no new permits for mineral exploration or licenses for exploitation would be approved by his government. This policy, also adopted by the two following administrations, amounted to a de facto moratorium on mining in the country. In response, a Canadian mining corporation, Pacific Rim, and a US mining corporation, Commerce Group, brought lawsuits against El Salvador in the World Bank's International Center for Settlement of Investment Disputes (ICSID), accusing the Salvadoran government of violating its own investment law and investor protection clauses of the Central American Free Trade Agreement.

Lawsuits have become increasingly common, according to a recent report by the Institute for Policy Studies, Mining Watch Canada, and the Center for International Environmental Law (Moore and Rocha 2019). The report analyzes a growing trend of multimillion-dollar claims brought by the mining industry against Latin American governments struggling to defend the environment and, in particular, freshwater resources. In 2009, however, this tactic was a new and powerful threat to the anti-mining movement of El Salvador. The country eventually won both cases, though the Pacific Rim case lasted seven years. The cases cost the Salvadoran state over eight million dollars in legal fees, monies that were urgently needed for healthcare, education, public security, and other priorities. With the finalization of the ICSID ruling, Pacific Rim and its new owner, OceanaGold, were ordered to reimburse this amount to the state. OceanaGold finally complied under continuous pressure from anti-mining advocates.

Phase 4 (2016–2017) saw the final and definitive battles for prohibition. With the Archdiocese of San Salvador and UCA taking the lead, the Catholic Church of El Salvador assumed a central role. This phase culminated in a ban on metallic mining in the country. The particular role that Catholic actors played in this process is detailed below.

Components of success

The national struggle against metallic mining in El Salvador was launched in highly unfavorable circumstances. El Salvador was governed at the time by a right-wing pro-business political party that prioritized direct foreign investment in its strategies for economic growth. The communities most threatened by mining had little experience with this industry and knew little about the costs to the environment and traditional livelihoods. The Salvadoran population in general had little awareness of the importance of natural resources, especially water, for the future viability of the country and was being aggressively bombarded by propaganda campaigns of transnational mining corporations highlighting the purported benefits of metallic mining.

Nevertheless, the struggle took root and grew, transforming itself into a national movement with the slogan "Yes to Life, No to Mining." Small-scale organizational and educational initiatives in local communities grew into a broad-based national campaign for legislative reform which eventually accumulated sufficient influence to shift the balance of power in Salvadoran society and force policymakers to prioritize national interests. The strategic vision that guided the struggle included several key components that distinguished the process from other anti-mining battles in the region, and contributed in important ways to the final victory.

Nonviolence

El Salvador is a country with a long and tragic history of violence, beginning with the Spanish conquest and followed by over three hundred years of brutal colonization, fifty years of repressive military dictatorship, and twelve years of bloody civil war. The country remains highly polarized in social, economic, and political terms, and highly susceptible to violence in the face of continuing social inequities and unresolved issues at the local and national levels. It was not surprising, therefore, when pro-mining interests in the country turned violent in 2009, resulting in the assassination of five anti-mining activists from the provinces of Cabañas, where the Canadian mining company Pacific Rim was insisting, at all costs, on mining gold and silver.

Pacific Rim also mobilized its workers and a group of local activists in aggressive street demonstrations against the Archdiocese of San Salvador, Oxfam America, and the Office of the Human Rights Ombudsperson, all of which opposed metallic mining. Within this highly volatile context, however, the organizers and key activists of the struggle in El Salvador insisted that it be one of ideas, not brute force; they believed in the power of truth, of objective scientific data, to overcome the aggression and well-financed propaganda campaigns of the transnational mining corporations.

Linkage between local communities and national movements

A second important feature was linking local community resistance to national advocacy campaigns, recognizing that the struggles of the poor at the local level are rarely sustainable if they are isolated from national processes. At the same time, it was understood that national campaigns are less viable and meaningful when they fail to incorporate the aspirations, motivations, and determined social force of broad-based community participation.

The methodology of public policy advocacy

A third key feature was the application of public policy advocacy methodologies. These non-violent processes were designed to influence policymakers via clear and precise proposals. This meant that the struggle could not be limited to street protests but had to include a clear policy agenda and well-planned advocacy campaigns designed to move that agenda forward. It was a disciplined effort by an organized citizenry to influence the formulation, approval, and implementation of public policies, programs, and practices through persuasion and social pressure.

A logic of empowerment

The struggle in El Salvador also focused on empowerment. Since public policy advocacy was considered to be an exercise of citizen power, advancing a policy agenda of prohibition required the empowering of key actors, especially communities most threatened by metallic mining. Far from being an abstract concept, empowerment meant organizing and educating communities and civil society at large and building alliances to enhance their capacity to influence legislators. It meant building and preserving unity; preparing strong, democratic, and well-informed leadership; assuring gender equity; building technical skills to access, analyze, and utilize information; generating knowledge through research; and building capacity for formulating viable policy goals and for developing the planning and negotiating skills essential to attaining policy victories.

A variety of strategies and tactics

A fifth feature of the struggle was the application of a wide variety of strategies and tactics. Among these were organizing and alliance-building, education and awareness-raising, research, lobbying, communications and media, and mobilization.

In practice, this meant traveling to distant communities to provide workshops, inform leadership, and organize. It meant building alliances at the national and international levels to include communities and key actors from El Salvador, other Central American nations, and countries that were home to transnational mining companies, especially the United States, Canada, and Australia. It meant holding public forums to inform citizens from all sectors of society, developing and disseminating educational materials, conducting public opinion polls and disseminating the results, lobbying government (including the presidency) and policymakers, working with the media to educate and inform public opinion, promoting popular consultations in municipalities threatened by mining, accompanying legislators on visits to former mining sites where environmental destruction from metallic mining continues to destroy freshwater resources and traditional livelihoods, and, of course, taking to the streets in periodic demonstrations.

The special role of the Catholic Church

It should be no surprise that the Catholic Church of El Salvador played a key role in the struggle. Since the early 1970s, Vatican documents have spoken of how unregulated human activity has caused a planetary ecological crisis. In his 1971 apostolic letter, *Octogesima Adveniens*, Pope St. Paul VI (1971, §21) referred briefly to the environmental challenges of the time, pointing out that:

> Man is suddenly becoming aware that by an ill-considered exploitation of nature he risks destroying it and becoming in his turn the victim of this degradation ... This is a wide-ranging social problem which

> concerns the entire human family. The Christian must turn to these new perceptions in order to take on responsibility, together with the rest of men, for a destiny which from now on is shared by all.

Pope St. John Paul II touched on this theme in 1979 in *Redemptor Hominis*. He called attention to worsening environmental problems and the need for conversion to remedy these, which would require enormous changes in lifestyles, models of production, and consumption (§15–16). In 2007, Pope Benedict XVI called for an end to the structural causes of environmental decline, including predominant models for economic growth that have shown themselves to be a threat to the environment and to life itself. And in 2015, Pope Francis presented his powerful encyclical, *Laudato Si'*, the first encyclical in the history of the Catholic Church to focus exclusively on the environment. In this document, Francis calls on the world's population to care for our common home by defending the environment and curtailing the abuse of natural resources, especially water. Like his predecessors, he criticizes systemic causes of environmental deterioration, pointing specifically to unbridled capitalism, a system that prioritizes capital over human life, generating inequity and imbalance between North and South and placing life itself at risk. Finally, he calls for a new ecological culture, with a different perspective and with policies and educational programs that lead to a new spirituality and a different way of life.

Many Catholics in El Salvador—Saint Oscar Romero, Father Rutilio Grande, the 1989 UCA martyrs, and many others—have been killed for actively living in solidarity with the poor and accompanying them in their struggles for social justice. Because of this history, the Catholic Church has enormous credibility and influence within Salvadoran society. It was a natural ally in the struggle against metallic mining because that struggle was grounded in the principles of Catholic social teaching, including respect for human dignity, respect for human life, free association, citizen participation, a preferential option for the poor and vulnerable, solidarity, and stewardship.

As early as the year 2000, Bishop Eduardo Alas Alfaro of Chalatenango opposed the incursion of transnational mining corporations in the communities of his diocese. In 2007, the Episcopal Conference of El Salvador formally pronounced its opposition to metallic mining with a document entitled "Cuidemos la Casa de Todos" ("Take Care of our Common Home"), arguing that:

> Our small country is the space in which God the Creator has called us to life. This is the portion of the world that He has entrusted to us to care for and to use according to His will ... But this blessed land that we dearly love suffers from growing and unmerciful deterioration. We are all responsible for conserving and defending it because the environment is the home of us all: of this and of future generations. (n.p.)

In February 2009, Monsignor José Luis Escobar Alas was installed as Archbishop of San Salvador. In his inaugural message, with elections upcoming and the country's president and the presidents of the legislature and judiciary all in attendance, the new Archbishop said: "To the departing government as well as the incoming government, whatever party you are from, I call upon you to prohibit the mining of precious metals. I ask this due to the grave damage it would cause to the health of our people through the contamination of our water" (Moreno 2009, n.p.). In the following years, Archbishop Escobar Alas has spoken out continually against metallic mining in press conferences, which are traditionally held after the main Sunday mass in the National Cathedral, and in other forums. His auxiliary bishop, Cardinal Gregorio Rosa Chavez, has done the same.

As the struggle grew in El Salvador, the persistent messaging of Pope Francis in defense of the environment motivated and strengthened the resolve of the Salvadoran Catholic bishops, priests, nuns, and laity to take on this cause with increasing emphasis and determination. Pope Francis called upon his church to be in the street, dirtied and tattered by its accompaniment of traditionally marginalized peoples. This call had an enormous impact on the hierarchy of El Salvador.

The commitment and role of the Salvadoran Church took a strategic leap in 2016 with the formulation of a new proposal for prohibition developed by legal experts and environmental specialists at UCA. It was not the first bill to be presented to the National Assembly: MESA presented one in 2006 and another was presented in 2013, but right-wing parties representing the interests of big business and closely linked to transnational corporate interests refused to debate either bill. The UCA proposal was shared with Archbishop Escobar Alas, who immediately gave it his support, and on February 6, 2017, UCA authorities presented it to the Commission on Environment and Climate Change of the Legislative Assembly, Archbishop Escobar Alas, Cardinal Rosa Chavez, and representation from the General Office for Justice, Peace and Integrity of Creation of the Franciscan Order.

By early March, 2017, the bill had barely advanced in the Assembly commission, so Archbishop Escobar Alas and other Catholic leaders called for a massive demonstration on March 9 (Velásquez 2017). Over six thousand priests, nuns, and laypeople, together with other social movements, responded to the call and marched from a park in the center of the capital city to the Legislative Assembly. There, in the presence of national and international media, we were met by representatives of each of the political parties. During the meeting that ensued, the president of the Legislative Assembly promised that the new bill would be approved before Easter Sunday, which was still several weeks away. Given the accumulated frustrations of seventeen years of organizing and struggle, few of the participants left the meeting convinced that the promise would be kept.

In the days that followed, transnational mining corporations like OceanaGold began to perceive that the balance of power was shifting.

Recognizing the growing threat to their interests, they returned to their earlier strategies of trying to win hearts and minds through aggressive media campaigns filled with promises of new technologies which would be more harmonious with the environment and would make "responsible mining," as they called it, possible. When they chose as their model their gold and silver mine in the province of Nueva Vizcaya, Philippines, little did they know how ready the experts of UCA were to rebut that claim.

Fully aware of the environmental destruction and human rights violations associated with gold mining in the Philippines, UCA invited the governor of Nueva Vizcaya to visit El Salvador and planned a week of public forums and meetings with communities, the President of El Salvador and his cabinet, and, most importantly, the Legislative Assembly´s Commission on Environment and Climate Change, which was also being intensely lobbied by OceanaGold. Using visual evidence of the destruction caused by OceanaGold in his province, the governor made convincing arguments against metallic mining. His 26 years as a political leader in his own country gave him great credibility with Salvadoran legislators and, in particular, the members of the Commission on Environment and Climate Change. Already highly knowledgeable about metallic mining as a result of receiving educational materials and direct testimony from UCA and other social actors over the years, the Commission members took the governor's presentation as a closing argument on the issue and voted to approve the Church/UCA proposal for the prohibition of metallic mining. On the following day, March 29, the bill was sent to the plenary session of the Legislative Assembly, where it was approved without opposition.

Conclusion

With the approval of this bill, El Salvador became the first country in the world to ban this controversial industry. Suffering from an abundance of negative superlatives (most violent nation, most densely populated, most deteriorated environment, most deforested, most water-stressed) this small nation became a precedent-setting example of citizen resolve, Catholic social action, national pride, and environmental protection, giving hope to communities and countries around the world who are confronting the abuses and violations of transnational mining corporations.

The long and difficult struggle to attain this once unthinkable goal was a gradual process consisting of several accumulating factors in which the Catholic community played a role:

- Social force bred by constructing and consolidating a social movement and empowering key social actors to analyze the threat of metallic mining, identify policy solutions, design specific policy proposals, and advance their political agenda through public policy advocacy;

- Knowledge about the vulnerable situation of El Salvador´s environment and the threat generated by metallic mining;
- Power to influence policymakers via lobbying, disseminating educational materials, and organizing and mobilizing increasingly broad sectors of society;
- Alliances at the local, national, and international levels that enlightened public opinion via effective communication and media strategies;
- Broad-based consensus among different sectors of society through dialogue with opponents and prioritizing national interests over individual and party interests; and
- Political will among decision makers to defend the environment.

The victory against metallic mining in El Salvador surprised and angered transnational companies and left them determined to dismantle this historic policy victory. The failure of the Salvadoran government to fully comply with the law two years after its approval is another serious problem. According to the law, by March 2019, small-scale artisanal mining, given a two-year grace period while communities searched for alternative livelihoods, should have been curtailed. Inactive mines around the country should have been adequately closed and environmental damage caused by mining activities in earlier decades should have begun to be remedied. None of this has occurred.

Nevertheless, public opinion and the will of lawmakers remain staunchly in favor of prohibition. More importantly, the struggle itself, with its emphasis on citizen empowerment and the application of methodologies of public policy advocacy instead of brute force, has offered a new way of doing politics in El Salvador that will hopefully consolidate this victory, strengthen an incipient democracy, transform traditional power relationships within society, and contribute to more profound and sustainable social change in El Salvador.

References

Agencia EFE. 2016. "La vida en El Salvador será inviable en 80 años por la crisis de agua, dice un estudio." May 12, 2016. https://www.efe.com/efe/america/sociedad/la-vida-en-el-salvador-sera-inviable-80-anos-por-crisis-de-agua-dice-un-estudio/20000013-2923438.

Agencia EFE. 2019. "El 80% del territorio salvadoreño estará en 'estrés hídrico' en el año 2022." March 22, 2019. https://www.efe.com/efe/america/sociedad/el-80-del-territorio-salvadoreno-estara-en-estreshidrico-ano-2022/20000013-3932816.

Benedict XVI. 2007. Letter to the Ecumenical Patriarch of Constantinople on the Occasion of the Seventh Symposium of the Religion, Science and the Environment Movement. September 1, 2007. http://www.vatican.va/content/benedict-xvi/en/letters/2007/documents/hf_ben-xvi_let_20070901_symposium-environment.html.

Calderón, Beatriz. 2019. "Poco favorables perspectivas hidrológicas para El Salvador: menos agua subterránea y disminución en caudales de ríos en próximos tres meses." *La Prensa Grafica*, May 2, 2019. http://www.laprensagrafica.com/elsalvador/Poco-favorables-perspectivas-hidrologicas-para-El-Salvador-menos-agua-subterranea-y-disminucion-en-caudales-de-rios-en-proximos-tres-meses-20190502-0295.html.

John Paul II. 1979. *Redemptor Hominis*. http://www.vatican.va/content/john-paul-ii/en/encyclicals/documents/hf_jp-ii_enc_04031979_redemptor-hominis.html.

La Prensa Gráfica. 2016. "90% de aguas superficiales del país están contaminadas." March 21, 2016. https://www.laprensagrafica.com/elsalvador/90--de-aguas-superficiales-del-pais-estan-contaminadas-20160321-0079.html.

Machuca, Evelyn. 2019. "45 pozos de la ANDA en el AMSS dejaron de producir agua entre 2008 y 2016." *La Prensa Grafica*, March 22, 2019. https://www.laprensagrafica.com/elsalvador/45-pozos-de-la-ANDAen-el-AMSS-dejaron-de-producir-agua-entre-2008-y-2016–20190321-0550.html.

McKinley, Andrés. 2016. *La Amenaza de La Minería Metálica En Un Mundo Con Sed*. San Salvador: UCA Editores. https://www.uca.edu.sv/iudop/wp-content/uploads/La-Amenaza-de-la-Mineri%c3%aca-Meta%c3%aclica-en-un-Mundo-con-Sed.pdf.

Moore, Jen, and Manuel Perez Rocha. 2019. "Extraction Casino: Mining Companies Gambling with Latin American Lives and Sovereignty through Supranational Arbitration." *Mining Watch Canada*, Institute for Policy Studies, and Center for International Environmental Law. https://miningwatch.ca/sites/default/files/isds_report_final_.pdf.

Moreno, Nestor Ivan. 2009. "Una Nueva Luz, un Nuevo Arzobispo…y el Pueblo De Dios." *La Voz de los Sin Voz*. February 20, 2009. https://pueblodedios.wordpress.com/2009/02/.

Paul VI. 1971. *Octogesima Adveniens*. http://www.vatican.va/content/paul-vi/en/apost_letters/documents/hf_p-vi_apl_19710514_octogesima-adveniens.html.

Perlez, Jane, and Kirk Johnson. 2005. "Behind Gold's Glitter: Torn Lands and Pointed Questions." *New York Times*, October 24, 2005. Accessed from Global Policy Forum, April 21, 2021. http://www.globalpolicy.org/socecon/tncs/2005/1024ring.htm.

Rodriguez, Roberto, Cristobal Garcia, and Efrain Castillo. 2007. "El Carácter Intermitente y Permanente Del Drenaje Ácido de Minas (AMD) En Zonas Semiáridas." *Ingenieria* 2, no. 2: 33–36.

University Public Opinion Institute. 2015. "Opinions and Perceptions towards Metals Mining in El Salvador." *University of Central America – José Simeón Cañas Newsletter* 29, no. 2. https://www.uca.edu.sv/iudop/wp-content/uploads/Press-Release.pdf.

Velásquez, Eugenia. 2017. "Ley contra la minería con respaldo en la Asamblea Legislativa." *ElSalvador.com*. March 9, 2017. https://historico.elsalvador.com/historico/317744/ley-contra-la-mineria-con-respaldo-en-la-asamblea-legislativa.html.

Section II

Mining and peace in Catholic theology and ethics

8 A just mining framework for the ethics of extraction of natural resources and integral peace

Tobias Winright

Introduction: from law enforcement and the use of force to mining

Although I am a Catholic ethicist and theologian, I previously worked as a law enforcement officer, in corrections and policing, and much of my scholarly and popular writing addresses the ethics of the use of force (Winright 2020). This book considers mining and conflict. A synonym for mining is *extracting*, which evokes memories of when I occasionally was part of an *extraction team*, tasked with *extracting* an obstinate, bellicose inmate from his cell in a maximum-security jail. If the team's presence at the cell's doorway and our verbal directions failed to persuade him to come out peacefully, then we had to enter the cell and escort him out. Sometimes doing so required the use of force, depending on the inmate's level of resistance: it might have sufficed to handcuff him and hold his arm as we accompanied him from the cell, or it might have entailed carrying him out while he was restrained in a wheelchair—and if the prisoner attempted to punch, kick, or bite us, then self-defense techniques would be employed, such as controlling holds involving pressure points. "Extraction" is "the action of taking out something (especially using effort or force)" (Vocabulary.com n.d.). The words "taking out something" reminds me, moreover, of when police refer to "taking out someone" (that is, subduing or shooting them) when, for instance, a suspect is on the verge of murdering an innocent hostage. Other synonyms for "extracting" include: removing, seizing, dislodging, pulling, coercing, prying, compelling, uprooting, wresting, wringing, displacing, and mining. These all involve the use of force, which is often associated with *conflict*.

Not all conflict, of course, involves the use of armed force or even violence. After all, as John Paul Lederach (2003) observes, "conflict is normal in human relationships, and conflict is a motor of change" (5); it is, in other similarly positive words, "a natural phenomenon that creates potential for constructive growth" (15). Still, conflict often holds a more negative connotation—that is, when it pertains to the use of force,

DOI: 10.4324/9781003094272-8

violence, war, destruction, and death. In such conflict, the use of force, whether in self-defense, policing, or war, bears a burden of proof before it can be regarded as morally just. In the words of the United States Catholic bishops, in their 1983 pastoral letter *The Challenge of Peace*, our starting point is a "presumption against the use of force" (§120). In the Catholic theological tradition, this moral stance or orientation concerning the use of force gave rise to the development of just war theory.[1] As the primary framework for moral reasoning about the use of force, just war theory traditionally has included two categories of criteria or principles: *jus ad bellum* includes criteria that justify going to war; and *jus in bello* includes principles for the just conduct of war. In recent years, a third category has also gained traction: *jus post bellum* includes principles and practices for the promotion and protection of a just and lasting peace after a war ends (Allman and Winright 2010).

Writing about natural resource extraction, which "endangers both the environment and peace," and the need for "sustainable mining" in South Africa, Peter Knox (2015) admits that this entire enterprise is "a concept in search of a theory" (117). Similarly, in an article on "The Missing Ethics of Mining," Shefa Siegel (2013) laments that "we are ready to discuss almost any other ethics before the ethics of mining," and asserts that "we have more faith in our capacity to restrain or end violence and war than to address the ethics of mining" (n.p.). Although Siegel seems correct about the amount of attention given to the ethics of violence and war in comparison to the ethics of mining, I question whether such deliberation necessarily must be conducted in an either/or, uncoupled fashion. Indeed, to examine extraction ethics, I suggest that we do not have to start from scratch: that the ethics of war may be applicable in an analogous way to the ethics of extraction, since both these activities involve the use of force. William P. George (2019), in his book *Mining Morality: Prospecting for Ethics in a Wounded World*, similarly suggests—without pursuing and developing it himself—that "the ethics of mining and the ethics of war, especially nuclear war, might be joined" (18). This is the task that I pursue in this chapter.

In what follows, I offer an exercise in what Lederach (2005) calls "moral imagination," a creative effort "to perceive things beyond and at a deeper level than what initially meets the eye" (26–27). Beneath the surface, the ethics of extraction, I reckon, is related to the ethics of the use of force as exhibited in just war theory. In the words of Lloyd Steffen (2012), there is "a way of ethical thinking," an "ethic that lies *behind* just war" that should be "widely applicable to all kinds of ethical issues," including but going "*beyond* the particular question of war or the use of force" (15, italics original). Although Steffen does not consider it, I suggest that this is the case for the morality of mining. Over the years, I have similarly addressed moral issues in law enforcement, inspired by Edward A. Malloy's (1982) invitation to ethicists to undertake an imaginative "exercise of analogical interpretation" using the reasoning and principles of just war theory to address police use of force (24). Accordingly, in this chapter I propose an *ethics of extraction* that includes a *just mining theory* with moral criteria encompassed under three categories: *jus ad extractionem*, *jus in extractione*, and *jus post extractionem*. This ethics of extraction, moreover, is meant to speak accessibly to government and corporate decision-makers, regardless of whether they share the Catholic faith and theological beliefs of the contributors to this volume. In doing so, it exemplifies what James M. Gustafson (1988, 1996), in his identification of four modes of moral discourse, refers to as ethical and policy discourse, rather than prophetic and narrative discourse. At the same time, I believe this proposal is coherent with a Catholic "integral peace" approach to addressing mining and conflict.

The chapter proceeds in four steps. In the first, the link between mining and conflict, between extraction and the use of force, is substantiated through attention to several cases. In the second, the ethic behind just war, which informs and suffuses just war thinking, is articulated as well as illustrated through its application to other moral issues. In the third, an ethics of extraction is constructed with a just mining theory comprised of three categories of criteria. In the fourth and concluding section, Gustafson's moral modes of ethical discourse are delineated to demonstrate how this chapter contributes to the development of a Catholic integral peace approach to addressing mining and conflict.

Mining and conflict: the use of force in mining

There are multiple ways mining and conflict are related. One under consideration within this present volume is the frequent correlation between mining and armed conflict. The so-called "resource wars," which are a part of wider "environmental conflict," occur between, or within, nations about scarce, finite natural resources (Detraz 2015). Mining sometimes is a *driver* of armed conflict, or perhaps more specifically, asymmetric natural resource wars. Rigobert Minani's chapter in this book describes how mining of materials like coltan and gold in the DRC is connected to violent conflict and profits for rebel groups (see also Koch and Kinsbergen 2018). Indeed, rebel groups, upon acquiring access to these resources within a conflict zone, are more able to finance their efforts against government military and police forces (Lujala 2010; Lee 2018). Similarly, Karl Gaspar's chapter notes how armed groups in the Philippines profit from mining indirectly by extorting mining companies. Of course, on the other hand, military, police, and paramilitaries, as well as mining corporations' own security forces, sometimes use force and employ violence not only against rebel groups but also indigenous communities. In Honduras, Bishop Santos Villeda (2018) describes how, through their funding of the government, foreign gold mining companies "subdue people by means of the armed forces, destroying churches and entire villages, even removing the dead from cemeteries in order to obtain the gold found in those areas" (93). In South Africa, labor disputes can lead to the loss of life, such as when police shot and killed thirty-four striking miners on August 16, 2012, at Lonmin's platinum mine in the town of Marikana (Knox 2015, 125). Andrés McKinley describes "aggressive street demonstrations" led by the Canadian mining company Pacific Rim against the Archbishop of San Salvador and others seeking to ban metallic mining in El Salvador at about the same time that five anti-mining activists were assassinated in Cabañas, where Pacific Rim was insisting, "at all costs," on mining gold and silver. And in this book, the chapters on Peru by Derechos Humanos y Medio Ambiente—Puno, and on Colombia by Héctor Fabio Henao and Sandra Polanía-Reyes, note instances of human rights and environmental defenders being targeted for intimidation and assassination.

In addition to actual armed conflict, the relationship between mining and conflict surfaces when we consider the negative, harmful consequences of mineral extraction itself on the environment and on people and communities. Put differently, mining inflicts violence on nature as well as on humans who are involved with, or impacted by, mineral extraction activities. As one environmental scientist notes, life-cycle assessments of metals that are commonly used in jewelry and technology reveal that gold and the platinum-group metals yield the greatest environmental burdens among metals, "as measured by cumulative energy use, global warming potential, human health implications, and ecosystem damage" (Klimas 2018, 211).

In South Africa, for example, mining operations have been implicated in acid water seepage, contaminating aquifers and rivers, as well as the release of airborne pollutants, such as asbestos dust, near population centers, causing lung cancer, asbestiosis, and mesothelioma (Knox 2015, 123). Similar harmful impacts to both the environment and people have been noted in the DRC, which holds three-quarters of the world's cobalt reserves and ranks first in industrial diamonds, second in iron, third in pyrochlore, fourth in copper, fifth in coltan, and eighth in cassiterite (Muhigirwa 2015, 35). Given the magnitude of mining activity there, Minani's chapter observes that environmental impacts in the DRC have been significant and varied, including deforestation, land degradation, and contamination of air and water. Likewise, in Peru, the gold industry has subjected miners to mercury exposure and poisoning, polluted waterways, and poisoned local communities who consume contaminated fish (Lemke 2016). In Honduras, Bishop Santos Villeda (2018) worries about the method of open-cut mining for gold, whereby foreign companies "use cyanide, which contaminates the ground, the air, and the water but gives them [these companies] great profits because it separates 97% of the gold particles" (92). Similarly, uranium mining in the United States exposed the Navajo people to radiation poisoning, including through contaminated drinking water that caused birth defects (Arnold 2014).

Some of these examples of harms both to people, especially indigenous and tribal peoples, and to the environment may be intentional. Indeed, some deleterious consequences appear to be direct and immediate. Others, though, may be the sort of "slow violence" identified by Rob Nixon (2011). These "ecological ripple effects," as Michael Schmitt (1997) calls them, do violence to and through ecosystemic relations. In this vein, Cynthia Moe-Lobeda (2013) details the structural violence of human institutions and systems, such as the inequitable distribution of power and privilege, that degrades, dehumanizes, injures, and kills persons by limiting or preventing their access to life necessities (72–78). As the Catholic Bishops of Appalachia (1975) observed, efforts to address environmental degradation in poor mining communities involve "a struggle against violence—against institutional violence—which sometimes subtly, sometimes brutally, attacks human dignity and life" (21). Indeed, exploitation and corruption are frequently associated with mining companies, especially foreign ones, and governments.

Significantly, conflictual, martial language (e.g., "violence," "attacks") is frequently employed by activists and religious leaders to refer critically to the harms resulting from mining, which brings us back to how *extracting* minerals invariably involves forcibly removing them and imposing negative consequences on the environment and on people. After detailing the harmful effects of gold mining on Mt. Diwata in the southern Philippines, including the mine tailings containing mercury that poison the Agusan River, along with the fish in these waters and the people that eat them, Randy J.C. Odchigue (2018) writes, "Taken together, these risks present *a clear and present danger* to the fragile ecosystem and the human community of the Agusan Marsh" (254, italics mine). In other words, mining there poses a grave and imminent threat both to the environment and to people. Similarly, in Brazil, Peter Hughes (2018) speaks of the need "to protect the rainforest" and the "urgent shared responsibility to protect life on the planet from increased global warming," even as he confesses "that we are losing the battle" (98–99). In this volume, Gaspar's chapter cites a 2019 document from the Philippine bishops that describes mining as a threat to the survival of the rural poor. Such wording (e.g., "clear and present danger," "to protect," "losing the battle," "threat to survival") further indicates the link between mining and conflict.

Pope Francis, as did his predecessors John Paul II (1990) and Benedict XVI (2010), also often makes use of martial language to describe the current climate and related environmental crises. In his homily on the feast of Saint Francis on October 4, 2013, Pope Francis (2013b) emphasized that God created our world "to be a place where harmony and peace can flourish," entailing a moral duty for us humans to "respect creation" and to "respect each human being" instead of being "instruments of destruction" of the environment and of human persons in "armed conflicts which cover the earth with blood" (n.p.). Likewise, in his first apostolic exhortation, *Evangelii Gaudium*, Pope Francis (2013a) expressed concern for vulnerable people whose lives are at risk and for "other weak and defenceless beings"—indeed, "creation as a whole"—threatened by destructive human activities motivated by "economic interests or indiscriminate exploitation" (§215). Although it is not explicitly mentioned, Pope Francis's lamentations about species extinction and desertification resulting from such harmful activities obviously echoes the above complaints about the baleful effects of natural resource extraction.

Moreover, in *Laudato Si'*, Francis (2015) emphasizes the link between conflict, or violence, and the harmful consequences for both planet and people. "The violence present in our hearts, wounded by sin," he writes, "is also reflected in the symptoms of sickness evident in the soil, in the water, in the air and in all forms of life" (§2). Also, since humans are not separate from, but are part of nature, the violence that we inflict upon nature is also an attack on ourselves, especially the poor and marginalized who are most vulnerable to the effects of environmental degradation (§48, 139).

In addition, the pope alludes to "resource wars" due to environmental degradation and natural resource shortages; "it is foreseeable," he warns, "that, once certain resources have been depleted, the scene will be set for new wars" (§57, 48). Not only do these actual conflicts harm both the environment and people, so too does ecological destruction impact the poor and "excluded" in ways that they are "treated merely as *collateral damage*" (§49, italics original). Here we see another invocation of terminology usually found in reference to conflict and war.

Mining is explicitly mentioned a handful of times in *Laudato Si'*. As in some of the examples earlier in Africa, Latin America, and the Philippines, the pope mentions that mercury pollution from gold mining and sulfur dioxide pollution from copper mining have "caused harm locally" (§51). He also notes how underground water sources are "threatened" by pollution that is produced, and rivers, lakes, and seas are contaminated by chemicals that are used, in "certain mining ... activities," especially in nations without sufficient governmental regulations (§29). Francis also decries the displacement of indigenous peoples and their communities "to make room for... mining projects which are undertaken without regard for the degradation of nature and culture" (§146). In each of these instances, the pope expresses concern for the environment and people, both of which are harmed by the activity of mining. His references to the *threats* posed by mining and its effects, including the *removal* of native populations from their homelands, call to mind, furthermore, the inextricable link between extraction and force.

Indeed, Francis makes precisely this point when he explicitly uses the word "extract" in his diagnosis of the conflictual relationship between humankind and nature.

> Men and women have constantly intervened in nature, but for a long time this meant being in tune with and respecting the possibilities offered by the things themselves. It was a matter of receiving what

> nature itself allowed, as if from its own hand. Now, by contrast, we are the ones to *lay our hands on* things, attempting to *extract* everything possible from them while frequently ignoring or forgetting the reality in front of us. Human beings and material objects no longer extend a friendly hand to one another; the relationship has become *confrontational.* (§106, italics mine).

The italicized words all could be used to describe uses of force. Francis clearly acknowledges that since ancient times humans have "intervened in nature," but he does not call for the absolute prohibition of such activities—as long as they are in harmony with, and respectful of, the environment. As the pope puts it, humans were receptive to what nature provided; they did not take or forcibly remove something from nature. His account, however, seems a bit out of sync with his initial use of the words "intervened in nature," unless perhaps by that he means something like "worked with" or "channeled," either of which would presumably be unconfrontational and nonforceful. Is the pope suggesting a complete return to the "in tune with" nature approach of intervening that was characteristic of earlier cultures and of indigenous populations today? If so, would this entail that mining should be abolished altogether?

To be sure, some activists and religious leaders seem to oppose mining absolutely, viewing it as inextricably harmful and violent, akin to how some also reject, if they are absolute pacifists, all other uses of force as immoral. For them, all mining, apparently, is deemed unjust. When pacifist Leo Tolstoy was asked about whether there is a difference between a terrorist's use of armed force and that by a police officer, he replied that the difference is as much as between cat and dog feces—adding that he doesn't like "the smell of either one or the other" (Simmons 1946, 651). So too does Hughes (2018) highlight "the withering verdict of Felipe Huamán Poma de Ayala, the first indigenous writer of the colonial period, who named mines and mining *el estiércol del diablo*: the devil's excrement!" (103). This absolutist approach is best exemplified in the leading role that the Catholic Church played in El Salvador's decision to ban metallic mining. However, given Francis's endorsement in *Laudato Si'* of some degree of intervention in nature, and his subsequent remarks at a meeting called "Mining for the Common Good" (2019), an absolute condemnation and prohibition of mining in all cases is not what he is calling for.

Still, as Knox (2015) posits, "It is difficult to think of any mining activity that actually enhances or even protects the integrity or biological diversity of natural systems" (124). Although the extraction of mineral resources provides beneficial goods, such as those that are part of the composition of my laptop computer that I am using right now, the process of mining itself, along with the litany of harmful consequences noted above, not only confronts but forces damage upon the immediate environment and the persons working there, and its deleterious effects ripple throughout the adjacent ecosystems and human communities. It also appears unlikely that mining can be conducted absent any harmful consequences at all. Knox observes, "Mining is an inherently dangerous activity" (125). It involves explosives, heavy machinery, and, as we have seen, industrial-strength chemicals. Although it may fail to enhance or protect the integrity of nature, mining provides people with goods—not only luxuries such as diamonds but also "needs" such as the cobalt that is used in lithium batteries and aircraft engine parts—while concurrently it damages or, at best, poses a risk of harm to the environment and to people. Plus, the goods that mining produces are inequitably distributed. To be sure, that our actions result in harms as well as benefits, evils as well as goods, is the case with many, if not all, human activities.

As Joseph A. Selling (2016) observes, there is always present in any human action an "interplay of good and evil," with the latter being not something neutral but "understood as anything that threatens, harms, or diminishes" human persons and, therefore, "should be avoided and minimized as much as possible" (191). Because I strongly doubt that there is going to be a full stop to mining in the foreseeable future, its harms cannot be completely avoided, but should be minimized. In the words of William George (2019), I am "asking not whether mining should be taken off the table, but rather how best to proceed in a manner that is not only economically and technologically feasible but also ecologically responsible and sustainable" (170). Thus, I propose an *ethics of extraction* that includes a *just mining theory* with justice as the thread running through moral criteria encompassed under three categories: *jus ad extractionem*, *jus in extractione*, and *jus post extractionem*. Before doing so, however, I need to explain a bit more the reasons for my use of terminology echoing that of just war theory.

An ethic behind and beyond just war

Standard approaches to ethics are usually classified as deontological, teleological (or utilitarian), or virtue theories. But there are also approaches to ethics that incorporate, or integrate, the strong points of these theories while recognizing their weaknesses. Steffen's (2012) work on a "hybrid ethic" that is "behind" and "beyond" just war thinking offers a helpful way of pursuing an ethics of extraction and a just mining theory. This ethic is grounded in the human capacity to reason and, therefore, is meaningful or persuasive not only to Catholics but to people of good will. Historically, it has been referred to as "natural law" (x).[2] As such, it is useful for public deliberation on moral issues, including the various forms of the use of force. Just war thinking is an expression and application of this hybrid ethic.

Importantly, this ethic is not meant to either rationalize or make easier the recourse to force, although that admittedly has often been the case through the misuse and abuse of just war theory by self-interested governments and nations; rather, it is intended to provide "a more vigorous public conversation where challenges are issued to—and through—the moral justification process itself" (Steffen 2012, xii). Steffen emphasizes that his effort to "extract the natural law-based and just war-related ethic" provides a framework for moral reflection and criteria for right action applicable to any issue, ranging from war to deception, from abortion to environmental protection (15, 17). Accordingly, the ethic offers "a system of guidance for deliberating, analyzing, and prescribing action that is good, right, and fitting," including "normative moral action guides" for gauging and measuring human activity (33, 37).

The initial premise of this hybrid ethic behind and beyond just war is that human persons affirm a just peace, which is not merely the absence of conflict or violence, which is a negative form of peace, but the more positive flourishing of harmonious relationships with one another and nature. In the Bible, the Hebrew word *shalom* refers to this just peace. *Shalom* is "a deep and all-embracing reality," which is not "simply the absence of violence," but a more positive and reconciling vision (Wirzba 2012, 66). Likewise, for Andean and Amazonian people in Bolivia, the indigenous concept of *vivir bien*, or "living well," means "a harmonic relationship between humans and nature" (Martins 2020, 131). Steffen (2012) presumes "as a given that reasonable persons of goodwill would prefer to resolve conflicts justly and without resorting to war or any use of force whatsoever if that is possible to do" (26). For Steffen, this is the "moral presumption" against the use of force (44–45), about which Lisa Sowle Cahill writes, "In my view, a presumption against violence … is necessary to Christian social ethics" (Cahill 2019, 31). Upon this fundamental premise, Steffen (2012) adds that "practical reasonableness necessarily imposes constraints on the use of force" which must be directed "to the rational and good end of peace" (43).

While Steffen (2012), like me, views just war thinking as applicable to issues beyond the use of force, he holds, as do I, that moral deliberations about "the prospective use of force inevitably involve the content of just war thinking" (21). On the relevance of just war reasoning to other moral issues, Steffen devotes attention to physician-assisted suicide, the withholding or withdrawing of medical treatment for severely disabled newborn infants, the death penalty and the criminal justice system, and abortion. Other ethicists have done so, too. In an essay on moral issues in biotechnology and genetics, Cahill (2005) examines the ethics of stem cell research, noting that "as in the case of many social issues, 'inaction' is not an option." Indeed, similarly to what I am calling for with mining, Cahill writes regarding embryonic stem cell research: "The task is therefore not to decide 'prospectively' whether it is a good idea, but to subject it to moral guidance and restraints, acknowledging its potential for beneficial outcomes while limiting the social damage and moral compromises it involves" (231). This looks quite similar, too, to what Steffen has been proposing. Thus, unsurprisingly, Cahill turns to just war thinking and to its criteria—which "can be used to discourage and restrain a practice without necessarily drawing an incontrovertible line against it"—to address the moral question of embryonic stem cell research (231, 232–35; see also Lauritzen 2001, 20–23).

Steffen (2012) also considers moral questions surrounding nonviolent resistance (e.g., boycotts, demonstrations, sit-ins) and how just war reasoning and criteria are implicitly exhibited in the thought and actions of Gandhi and Martin Luther King, Jr. Even their nonviolent resistance was a use of force, perhaps even coercive force, which required justification as well as measured application (51–72).[3] Although Steffen does not mention it, similar just-war analyses of nonviolent force were provided by the influential ethicists James Childress (1971) and Paul Ramsey (1961) over half a century ago. While not always seeing eye to eye on moral methods, Childress and Ramsey wrote extensively on questions both in just war theory and in bioethics (McCarty 2018). Less noted is their attention to nonviolent resistance and protest, and their agreement that just war thinking should be applied to such actions. A similar claim has been made by Maryann Cusimano Love (2010) about her "just peace principles," which she regards as resembling just war principles and as implied in just war thinking (56–57). For his part, Childress (1971) noted about civil disobedience:

> The "just war doctrine" offers a set of considerations for determining when war is justified, and analogous criteria must be employed in determining when civil disobedience is justified, although perhaps it is more accurate to suggest that civil disobedience is subject to the same general demands of morality as any other action rather than that it is illuminated by just war criteria. However that may be, certainly the appropriate criteria for evaluating civil disobedience coincide to a great extent with traditional just war criteria such as just cause, good motives and intentions, exhaustion of normal procedures for resolving disputes, reasonable prospect for success, due proportion between probable good and bad consequences, and right means. (204)

I think this insight concerning nonviolent force is significant. Cahill (2019), too, notes that both armed force and nonviolence involve moral dilemmas, ambiguities, and messiness. Not only are there moral costs that result from the use of violent force, so too "the renunciation of violence is not without its own human and moral price" (323). Decisions and actions—including when morally justifiable—"can still have *morally* evil dimensions, not just unfortunate and regrettable ones" (126, italics original). This is precisely the point made by Selling about the "interplay between good and evil" in human actions and the moral requirement that any evil ought to be avoided, if possible, or at least minimized.

Steffen (2012) also alludes to the applicability of the just war ethic to policing, an issue about which I have written extensively (41; Winright 2020). In most nations the police are authorized to use force, including lethal force, so an analogous application of the moral reasoning and criteria of just war theory to guide a police officer's decision to use force and to moderate its performance makes sense. As Irish moral theologian Enda McDonagh (1980) has written, "Accepting, in common with the majority of Christians past and present, the need for the violence of restraint in society, one is operating with criteria similar to those of the just war" (71).[4] That the just war ethic is most applicable to the use of armed, and even unarmed, force is a point that has been made consistently by a range of Catholic and Protestant ethicists. Hence, it makes sense to extend this ethic to the use of force in mining. It is to this exercise of moral imagination and analogical interpretation of just war thinking and criteria to the extraction of mineral resources that I shall now turn.

A just mining framework with action guiding criteria

Steffen (2012) labels the criteria of just war thinking "action guides ... not intended to rationalize war" but to "impose restraint, insisting that force be used only in a way that is proportionate to the end of addressing injustice and restoring peace" (40). These criteria, or principles, traditionally fall under two categories, *jus ad bellum* and *jus in bello*. In recent years, a third category, *jus post bellum*, has gained traction, although its criteria emerge from taking seriously the obligations of the criterion of right intent—that is, the establishment of a just peace—that serves as the unifying thread holding together all the categories and their criteria (Bell, Jr. 2009; Capizzi 2015; Allman and Winright 2010; 2012). Before constructing a set of analogous just mining criteria, these just war categories and principles need to be delineated.

Depending on the author, the number of criteria varies. Steffen (2012) identifies nine, and so do the US Catholic bishops (1983), although there are slight differences in their respective lists (22; §84–110). The *Catechism of the Catholic Church* (1993) refers to "the traditional elements enumerated in what is called the 'just war' doctrine," but compared to Steffen and the US bishops, it mentions half as many "strict conditions for legitimate defense by military force [which] require rigorous consideration" (no. 2309). I collate the criteria from these sources in what follows.

The first main set of criteria in just war theory is known as *jus ad bellum*, which includes several criteria, all of which ought to be satisfied prior to embarking upon war. These principles are concerned with *why* and *when* going to war is justified. The first criterion is just cause, and it deals specifically with the *why* aspect of going to war and using armed force. A just war is undertaken in response to serious injustice, such as an act of aggression or an attack. Both the *Catechism of the Catholic Church* and the

Compendium of the Social Doctrine of the Church underscore the right and the duty of nations to use force of arms to protect, under the rubric of "legitimate defense," their own citizens and innocent victims in other countries who are unable to defend themselves (Pontifical Council for Justice and Peace 2004, §500, 504). The second criterion is legitimate authority. Only duly appointed political authorities who bear responsibility for protecting the common good may declare and wage war. In modern democracies with representative governments, the decision to go to war does not fall on the shoulders of only one person. The third criterion, right intent, "means pursuit of peace and reconciliation, including avoiding unnecessarily destructive acts or imposing unreasonable conditions" (National Conference of Catholic Bishops 1983, §95). Ultimately, the goal is to restore order, which means setting in motion necessary conditions for establishing a just peace for all who are involved and impacted. This is the *telos* of just war, which thereby "requires placing just war theory within the larger framework of peacemaking" (Baer and Capizzi 2006, 170). On the fourth criterion, as the *Catechism* (1993) puts it, "there must be serious prospects of success." There must be a reasonable hope of winning the war and achieving the aim of restoring a just peace. The fifth criterion is last resort. As the *Catechism* stipulates, "all other means of putting an end to [conflict] must have been shown to be impractical or ineffective" (no. 2309). The sixth criterion, proportionality, weighs the evils associated with going to war; these evils must not be greater than the already-present evils that political authorities are seeking to prevent or stop. Proportionality assesses whether going to war will result in more harm, damage, and costs than any good that might be achieved.

The second major category of criteria for just war is *jus in bello*, which is concerned with *how* the war is conducted once hostilities have begun. These criteria gauge the *means* employed, including weapons, tactics, and targeting. These criteria restrain and limit the use of force. The first criterion within *jus in bello* is discrimination, or noncombatant immunity. Militaries are supposed to discriminate, or distinguish, between combatants and non-combatants. Moreover, certain places, such as hospitals, schools, and sacred buildings, should not be targeted. The other side of the coin, however, is that some unintentional injuries and deaths of civilians (and damage to non-military property) *may* be permitted and excused under the euphemism of "collateral damage." At play in this distinction is the principle of double effect, which recognizes that there are multiple effects—good and evil—from an action (Cavanaugh 2006). As long as the harmful consequences, although foreseen, are not intended, as long as these harmful consequences are not the means to the end that is sought, and as long as these harmful consequences do not outweigh the good results, then the use of force may be considered just. The second criterion within *jus in bello* is proportionality, which considers on a micro level the fittingness of the actual use of force during war rather than the more macro considerations associated with the war itself. Militaries must use only the amount of force necessary, and not excessive force, to achieve their objective. One ought not to use a sledgehammer to swat a mosquito on her neighbor's forehead. That would be excessive, disproportionate force.

The third category of criteria is *jus post bellum*, which provides a litmus test for the right intention of establishing a just peace for all at war's end (Allman and Winright 2010; Stahn and Kleffner 2008). Because this is a developing area in just war theory, the principles and expectations vary from author to author. For a state or military force that was justified in going to war and was just in its conduct during that war, justice still entails responsibilities after the shooting stops. These include restorative justice, compensation, security and public safety, economic recovery, and environmental cleanup. Such expectations are directed toward the establishment of a just peace, minimizing the likelihood of rekindled conflict.

What might it look like were we to convert or translate these just war criteria into just mining criteria? Again, what I have in mind is akin to the concrete "action guiding principles" of "protect, respect and remedy" that the United Nations Human Rights Office of the High Commissioner (2014) has provided for governments and companies to facilitate the development of consonant policies, rules, and processes. Like just war theory, just mining theory has criteria, or action guides, fitting under three categories: *jus ad extractionem*, *jus in extractione*, and *jus post extractionem*.

For the category of *jus ad extractionem*, which refers to the justified commencement of mining, the first criterion, just cause, points to the need for the minerals to be mined and the necessity of mining to acquire them. Because the beginning presumption is against the use of force, which includes the use of force in mining that harms the natural environment and human communities, the first criterion requires that the goods provided by the minerals must really be human needs that justify the use of force that harms the environment and risks the well-being of miners and nearby communities.[5] The theological basis for this initial presumption is articulated by Ferdinand Muhigirwa (2015), who writes, "As a gift of God, all creation and the environment must be safeguarded and protected against harm and adverse environmental impacts that affect the [sic] nature, air, water, forests, wildlife, and the ecosystem" (36). If anything, given the "clear and present danger to the fragile ecosystem and the human community," about which Odchigue (2018) warns, for example, "in the Agusan Marsh" (254), there is more just cause to defend ecosystems and human communities against the harmful threats posed by mining. Therefore, the bar is high for there to be just cause for mineral extraction. In addition, just cause entails that we ask the question about who really benefits from mining, because as Knox (2015) warns, echoes of colonialism and privilege would not constitute just causes for mining (121).

Regarding the second criterion within *jus ad extractionem*, just authority, part of the problem with mining, as noted by Muhigirwa (2015), is the "weakness of the state's regulatory powers" (36). Odchigue (2018) similarly notes that in the Philippines "there appears to be a failure of government agencies to implement regulatory policies for small- and large-scale mining" (255). Knox (2015) adds that the state has a duty "to protect private ownership, which is a human right, but at the same time to ensure that the exercise of this right does not force some to remain in poverty" (122). And in this book, Elias Opongo analyzes the way poor regulation of mining in Eastern Africa can specifically impact conflict by failing to engender sustainable national development. All of these words of caution and recommendation imply more democratic and representative participation in the decision-making processes related to mining—including governments and local community members—so that mining corporations, with their governing boards and executives, are not making decisions alone that impact the lives of others.

The third criterion, right intent, is supposed to be the establishment of a just peace for all. The US bishops (1983) extend this criterion to encompass nature, too: "[T]rue peace implie[s] the restoration of the right order not just among peoples, but within all of creation" (§32). As Santos Villeda (2018) notes, drawing from the *Compendium of the Social Doctrine of the Church*, there is a fundamental rule about "the intervention of the human being in nature [which] should be governed by respect for other persons and their rights as well as respect toward other living creatures" (94; Pontifical Council for Justice and Peace 2004, §463). Such respect for all entails giving due attention to their safety over against the dangerous harms that may accompany and result from the force of mining. As Santos Villeda (2018) reminds us, security should not only be provided for the companies but also for the people (93). Right intent will, moreover, spill over into the other categories and criteria of just mining.

The fourth criterion, the probability of success, should take into consideration not only the likelihood of acquiring resources through mining, but also the prospects for doing so in a way that minimizes harm to the environment and people, as well as its odds of benefitting all involved, such as by creating integral human development and preserving cultural and environmental heritages. As such, this criterion is related to other criteria, such as proportionality, and serves as a reminder that for an action using force, such as mining, to be considered morally justified, all the criteria—not only one or some—must be satisfied.

The *jus ad extractionem* criterion of last resort asks whether other reasonable and feasible means of obtaining mineral resources have been first attempted and exhausted. Or, have other alternative sources of what a particular mineral provides been seriously explored? Another way of getting at this criterion is to see it as connected to the ecological "precautionary principle," which says to err on the side of precaution, to make sure we have explored other ways of moving forward that avoid or minimize the negative effects of the present course of action that is under consideration

(Brinkman 2013, 207). Because the dangerous "risks [of mining] are all known," Knox (2015) advises that "reasonable precautions...[should] be in place" (125). One pertinent practice to consider here is recycling, which, as the chapter by Holden and Montevecchio notes, can greatly reduce the need for new mining even if it would mean lower corporate profits and higher consumer costs.

For the *jus ad extractionem* criterion of proportionality, the ecological concept of sustainability seems apt. On sustainability, Christiana Zenner and Andrea Vicini (2015) begin with the 1987 report of the Bruntdland Commission, *Our Common Future*, which linked the environmental and social impacts of economic development, and called for sustainable development to mean meeting people's needs today while protecting sufficient natural resources for future generations (1).[6] As they note, sustainability is related to "the norm of justice," which entails "fairness to future generations as well as respect for ecosystems and the earth processes on which all forms of life (including but not limited to humans) depend" (3). With regard to mining, Knox (2015) thinks the "concept of 'sustainable mining' presents an inherent paradox," since these natural resources are "neither in infinite supply nor constantly being renewed" (117). Weighing justice in these ways is reflective of using proportionality.

In the second category, *jus in extractione*, with its two criteria of discrimination and proportionality, the focus turns to the actual means employed in the mining operations. Discrimination would rule out direct, intentional attacks on that which is necessary for human life, such as water, food, and homes. As with surgery in medical ethics and with surgical strikes in just war ethics, extraction ethics expects that the intended damage from the use of force by mining be immediately directed to the area mined. That is, mining companies and workers must discriminate, or draw a distinction, between the plot of earth to be mined and the surrounding area. They should also use the most fitting, least harmful means of mining, thereby minimizing dangers and risks. The impacts on necessities such as water should also be taken into account in assessing proportionality. The fact that the national ban on mining in El Salvador was based on protecting water is an example of how the norm of discrimination and proportionality might apply. Moreover, any foreseeable harms to the wider ecological systems and human communities must be avoided or curtailed as much as possible. Under these circumstances any collateral damage to the wider environment and population might perhaps be excusable, employing the principle of double effect, as long as the harmful consequences are not disproportionate to the benefits from the mining for the company, the consumer elsewhere, and the community there. These criteria for *jus in extractione* should also take into account the future considerations that arise with sustainability. There should also be just remuneration and compensation for workers (Knox 2015, 125).

The third category of just mining criteria, *jus post extractionem*, requires that mining companies take responsibility for what they have done to the mining site and its environs, as well as the ongoing consequences of the mining conducted there. Santos Villeda (2018) urges that "it is necessary to evaluate the long-term environmental cost of extractive activities" (95). So too does Knox (2015) anticipate the "costly cleanup" of environmental damage (120). Not only should toxic waste be removed, perhaps trees and flora should be planted. For affected human communities, adequate health care and financial compensation should be provided and the land should be restored for new development activities or other uses. Post-mining expectations such as these help ensure that the pre-mining right intent has been sincerely implemented throughout all phases.

Conclusion: extraction ethics and integral peace

James M. Gustafson (1996) identified four types of moral discourse: ethical, policy, prophetic, and narrative. This chapter intentionally exemplifies the first two types. The task of the ethical mode is to provide guidance for deciding "how one ought to act in particular circumstances" by referring to concepts such as rights and duties, as well as norms like justice (39). The task of the policy mode is to assess practices, institutions, and action guides for persons within social institutions such as governments and corporations. This mode takes seriously limitations and conflicts within concrete situations as well as the need for compromise and accommodation. By attempting to communicate persuasively to business and governmental audiences, by giving attention to the interplay of goods and inevitable evils accompanying mining as a use of force, and by attempting to limit these evils while stopping short of the prohibition of mineral extraction altogether, my just mining framework and criteria also are consonant with the ethical and policy discourses.

Perhaps more absolute stances against the use of force—whether in war or mining—fit within Gustafson's other two types of moral discourse. The prophetic mode attempts to widen the scope of vision and reorient worldviews. As such, it tends to be "more general than ethical discourse" (Gustafson 1996, 41). It also employs narratives sometimes, especially from Scripture, to indict injustice and to inspire hope for a just peace. The narrative type serves to shape the character of persons and the ethos of communities. It tends to be less universal, its audience more distinctively religious. It also tends not to provide concrete answers to complex moral problems.

The hybrid ethic that I am suggesting can include all four of these types of moral discourse. Because this essay's primary audience is mining professionals and political authorities, I have mostly employed the ethical and policy discourses. That is not to say that this just mining framework cannot be useful for religious leaders and activists, including those who are more, if not absolutely, opposed to mining. For them, I would suggest that this just mining framework and its principles be viewed as what a number of

Mennonite ethicists have called "middle axioms," which call "other people or institutions to live up to their own stated moral principles, highest ideals, or long-term enlightened self-interests without endorsing all of the assumptions behind others' ethical systems" (Schlabach 2019, 219). The concrete action guides I have suggested can be appealed to by anyone to hold those who do mining accountable to justice.

Daniel Cowdin (2008) observes that Catholic environmental ethics has "difficulties...at the level of specific principles and norms...mediating between its deeper theological foundations and the specificity of concrete situations" (166). In his view, it "lacks the kind of action-guiding moral traction that is typical of other areas in the Catholic moral tradition," such as just war thinking and bioethics (180-81). This chapter has attempted to suggest these sort of concrete action guides for mining. Extraction ethics entails a just mining framework, consisting of criteria that serve as action guides prior to mining (*jus ad extractionem*), during mining (*jus in extractione*), and following mining (*jus post extractionem*). Adherence to this just mining framework, it is hoped, should mitigate the devastating consequences of mining to the environment and people. In doing so, just mining should, moreover, diminish the likelihood of causing or exacerbating conflict. It should also contribute to "integral peace" (Winright 2016; 2018). Throughout *Laudato Si'*, Pope Francis (2015) uses the word *integral*, especially when calling for the promotion of an "integral ecology" that recovers "a serene harmony with creation" and "break[s] with the logic of violence, exploitation and selfishness" (§225, 230). Francis also uses the term when referring to the need for contributions from an array of disciplines and sources, including science, economics, politics, ethics, and faith (§137). Here he seems to have in mind not only making connections and interrelating insights from these various sources, but *integration*, or synthesis (Butkus and Kolmes 2011). Integral peace, I suggest, is the end, or *telos*, of the hybrid ethic behind and beyond just war thinking, as exemplified in an extraction ethics consisting of a just mining framework.

Notes

1 Sometimes called "just war tradition" and "just war thinking," I regard "just war theory" as a part of the just war *tradition*, which includes multiple versions or theories, as well as developments over the years, and all of which are expressions of just war *thinking*. For more on the language of "presumption," see Winright (2009).

2 This just war mode of reasoning can be found in non-Western and non-Christian cultures as well (Lo and Twiss 2015; Kelsay 1993; 2007).

3 As Steffen acknowledges, this observation about the coerciveness and destructive consequences of nonviolent methods was made by prominent ethicist Reinhold Niebuhr (1932, 241).

4 This interpretation echoes that of Ralph B. Potter (1973, 49–50), Edward A. Malloy (1982, 10, 24), and Paul Ramsey (1968), who argued that the "moral economy" of the just war tradition is "*morally* if not *legally* binding upon the use

of force between nations," and it also "regulates the use of force within political communities, where it is both *morally* and *legally* binding" (144).

5 This concern is affirmed in this book by Douglass Cassel, who argues that the end-use of minerals should be a consideration in the ethical analysis of the operation to be undertaken.

6 The Brundtland Commission stated that "sustainable development is development that meets the needs of the present without compromising the ability of future generations to meet their own needs" (United Nations World Commission on Environment and Development 1987, ch. 2, §1).

References

Allman, Mark J., and Tobias L. Winright. 2010. *After the Smoke Clears: The Just War Tradition and Post War Justice*. Maryknoll, NY: Orbis Books.

Allman, Mark J., and Tobias L. Winright. 2012. "Growing Edges of Just War Theory: *Jus Ante Bellum*, *Jus Post Bellum*, and Imperfect Justice." *Journal of the Society of Christian Ethics* 32, no. 2: 173–191. 10.1353/sce.2012.0039.

Arnold, Carrie. 2014. "Once Upon a Mine: The Legacy of Uranium on the Navajo Nation." *Environmental Health Perspectives* 122, no. 2: A44–A49. 10.1289%2Fehp.122-A44.

Baer, Helmut David, and Joseph E. Capizzi. 2006. "Just War Theory and the Problem of International Politics: On the Central Role of Just Intention." *Journal of the Society of Christian Ethics* 26, no. 1: 163–175. https://www.jstor.org/stable/23561502.

Bell, Daniel M., Jr. 2009. *Just War as Christian Discipleship: Recentering the Tradition in the Church Rather Than the State*. Grand Rapids, MI: Brazos Press.

Benedict XVI. 2010. "If You Want to Cultivate Peace, Protect Creation." World Day of Peace message. January 1, 2010. http://w2.vatican.va/content/benedict-xvi/en/messages/peace/documents/hf_ben-xvi_mes_20091208_xliii-world-day-peace.html.

Benedict XVI. 2011. https://www.vatican.va/content/benedict-xvi/en/apost_exhortations/documents/hf_ben-xvi_exh_20111119_africae-munus.html.

Brinkman, John T. 2013. "Discernment of the Church and the Dynamics of the Climate Change Convention." In *Environmental Justice and Climate Change*, edited byJame Schaefer and Tobias Winright, 197–214. Lanham, MD: Lexington Books.

Butkus, Russel A., and Steven A. Kolmes. 2011. *Environmental Science and Theology in Dialogue*. Maryknoll, NY: Orbis Books.

Cahill, Lisa Sowle. 2005. *Theological Bioethics: Participation, Justice, and Change*. Washington, DC: Georgetown University Press.

Cahill, Lisa Sowle. 2019. *Blessed Are the Peacemakers: Pacifism, Just War, and Peacebuilding*. Minneapolis: Fortress Press. 10.2307/j.ctv9b2ww5.

Capizzi, Joseph E. 2015. *Politics, Justice, and War: Christian Governance and the Ethics of Warfare*. Oxford: Oxford University Press. 10.1093/acprof:oso/9780198723950.001.0001.

Catechism of the Catholic Church. 1993. Vatican City: Libreria Editrice Vaticana.

Catholic Bishops of Appalachia. 1975. *This Land Is Home to Me: A Pastoral Letter on Powerlessness in Appalachia by the Catholic Bishops of the Region*. Martin, KY: Catholic Committee of Appalachia.

Cavanaugh, T.A. 2006. *Double Effect Reasoning: Doing Good and Avoiding Evil.* New York: Oxford University Press. 10.1093/0199272190.001.0001.

Childress, James F. 1971. *Civil Disobedience and Political Obligation: A Study in Christian Social Ethics*. New Haven, CT: Yale University Press.

Cowden, Daniel. 2008. "Environmental Ethics." *Theological Studies* 69, no. 1: 164–184. 10.1177%2F004056390806900109.

Detraz, Nicole. 2015. *Environmental Security and Gender*. New York: Routledge. 10.4324/9781315765037.

Francis. 2013a. *Evangelii Gaudium*. http://w2.vatican.va/content/francesco/en/apost_exhortations/documents/papa-francesco_esortazione-ap_20131124_evangelii-gaudium.html.

Francis. 2013b. "Homily of the Holy Father Francis: Pastoral Visit to Assisi." October 4, 2013. http://www.vatican.va/holy_father/francesco/homilies/2013/documents/papa-francesco_20131004_omelia-visita-assisi_en.html.

Francis. 2015. *Laudato Si'*. http://www.vatican.va/content/francesco/en/encyclicals/documents/papa-francesco_20150524_enciclica-laudato-si.html.

Francis. 2019. "Address to Participants at the Meeting Promoted by the Dicastery for Promoting Integral Human Development on the Mining Industry." May 3, 2019. http://www.vatican.va/content/francesco/en/speeches/2019/may/documents/papa-francesco_20190503_incontro-industria-mineraria.html.

George, William P. 2019. *Mining Morality: Prospecting for Ethics in a Wounded World*. Lanham, MD: Lexington Books.

Gustafson, James M. 1988. *Varieties of Moral Discourse: Prophetic, Narrative, Ethical, and Policy*. Grand Rapids, MI: Calvin College.

Gustafson, James M. 1996. *Intersections: Science, Theology, and Ethics*. Cleveland, OH: Pilgrim Press.

Hughes, Peter. 2018. "The Pan Amazon, Extractive Industries, and the Church." In *Fragile World: Ecology and the Church*, edited by William T. Cavanaugh, 97–111. Eugene, OR: Cascade Books.

John Paul II. 1990. "Peace with God the Creator, Peace with All of Creation." World Day of Peace Message. January 1, 1990. http://www.vatican.va/holy_father/john_paul_ii/messages/peace/documents/hf_jp-ii_mes_19891208_xxiii-world-day-for-peace_en.html.

Kelsay, John. 1993. *Islam and War: A Study in Comparative Ethics*. Louisville, KY: Westminster/John Knox Press.

Kelsay, John. 2007. *Arguing the Just War in Islam*. Cambridge, MA: Harvard University Press. 10.2307/j.ctvjz81s7.

Klimas, Christie. 2018. "Christian Christmas Consumption: Ethical Considerations of the Environmental and Social Impacts of Holiday Spending." In *Fragile World: Ecology and the Church*, edited by William T. Cavanaugh, 210–218. Eugene, OR: Cascade Books.

Knox, Peter. 2015. "Sustainable Mining in South Africa: A Concept in Search of a Theory." In *Just Sustainability: Technology, Ecology, and Resource Extraction*, edited byChristiana Z. Peppard and Andrea Vicini, 117–130. Maryknoll, NY: Orbis Books. 10.1353/sce.2018.0018.

Koch, Dirk-Jan, and Sara Kinsbergen. 2018. "Exaggerating Unintended Effects? Competing Narratives on the Impact of Conflict Minerals Regulation." *Resources Policy* 57: 255–263. 10.1016/j.resourpol.2018.03.011.

Lauritzen, Paul. 2001. "Neither Person nor Property: Embryo Research and the Status of the Early Embryo." *America* 184, no. 10 (March 26, 2001): 20–23.

Lederach, John Paul. 2003. *Little Book of Conflict Transformation: Clear Articulation of the Guiding Principles by a Pioneer in the Field*. New York: Good Books.

Lederach, John Paul. 2005. *The Moral Imagination: The Art and Soul of Building Peace*. New York: Oxford University Press. 10.1093/0195174542.001.0001.

Lee, Chia-yi. 2018. "Oil and Terrorism: Uncovering the Mechanisms." *Journal of Conflict Resolution* 62, no. 5: 903–928. 10.1177%2F0022002716673702.

Lemke, Rebekah Kates. 2016. "The Pull of Peru's Gold Rush." CRS.org. December 30, 2016. https://www.crs.org/stories/peru-illegal-gold-mining-climate-change.

Lo, Ping-Cheung, and Sumner B. Twiss, eds. 2015. *Chinese Just War Ethics: Origin, Development, and Dissent*. London: Routledge. 10.4324/9781315740706.

Love, Maryann Cusimano. 2010. "What Kind of Peace Do We Seek? Emerging Norms of Peacebuilding in Key Political Institutions." In *Peacebuilding: Catholic Theology, Ethics, and Praxis*, edited byRobert J. Schreiter, R. Scott Appleby, and Gerard F. Powers, 56–91. Maryknoll, NY: Orbis Books.

Lujala, Päivi. 2010. "The Spoils of Nature: Armed Civil Conflict and Rebel Access to Natural Resources." *Journal of Peace Research* 47, no. 1: 14–28. 10.1177%2F0022343309350015.

Malloy, Edward A. 1982. *The Ethics of Law Enforcement and Criminal Punishment*. Washington, DC: University Press of America.

Martins, Alexandre A. 2020. *The Cry of the Poor: Liberation Ethics and Justice in Health Care*. Lanham, MD: Lexington Books.

McCarty, Brett. 2018. "Medicine as Just War? The Legacy of James Childress in Christian Ethics." *Journal of the Society of Christian Ethics* 38, no. 2: 57–74. 10.1353/sce.2018.0033.

McDonagh, Enda. 1980. *Church and Politics: From Theology to a Case History of Zimbabwe*. Notre Dame, IN: University of Notre Dame Press.

Moe-Lobeda, Cynthia. 2013. *Resisting Structural Evil: Love as Ecological-Economic Vocation*. Minneapolis: Fortress Press.

Muhigirwa Rusembuka, Ferdinand. 2015. "Theological Perspectives on Governance in the Mining Sector in the Democratic Republic of Congo." In *Just Sustainability: Technology, Ecology, and Resource Extraction*, edited by Christiana Z. Peppard and Andrea Vicini, 35–38. Maryknoll, NY: Orbis Books. 10.1353/sce.2018.0018.

National Conference of Catholic Bishops. 1983. *The Challenge of Peace: God's Promise, Our Response*. Washington, DC: United States Catholic Conference.

Niebuhr, Reinhold. 1932. *Moral Man and Immoral Society*. New York: Scribner.

Nixon, Rob. 2011. *Slow Violence and the Environmentalism of the Poor*. Cambridge, MA: Harvard University Press. 10.4159/harvard.9780674061194.

Odchigue, Randy J. C. 2018. "Nurturing Communities, Sustaining Fragile Ecologies." In *Fragile World: Ecology and the Church*, edited by William T. Cavanaugh, 252–270. Eugene, OR: Cascade Books.

Pontifical Council for Justice and Peace. 2004. *Compendium of the Social Doctrine of the Church*. Washington, DC: United States Conference of Catholic Bishops, 2004.

Potter, Ralph B. 1973. *War and Moral Discourse.* Richmond, VA: John Knox Press.

Ramsey, Paul. 1961. *Christian Ethics and the Sit-In.* New York: Association Press.

Ramsey, Paul. 1968. *The Just War: Force and Political Responsibility.* New York: Charles Scribner's Sons.

Santos Villeda, Luis Alfonso. 2018. "Extractive Industries, Destructive Industries: The Case of Honduras—A Fragile Country and Devastated Ecology." In *Fragile World: Ecology and the Church*, edited by William T. Cavanaugh, 91–96. Eugene, OR: Cascade Books.

Schlabach, Gerald W. 2019. *A Pilgrim People: Becoming a Catholic Peace Church.* Collegeville, MN: Liturgical Press.

Schmitt, Michael. 1997. "Green War: An Assessment of the Environmental Law of International Conflict." *Yale Journal of International Law* 22: 1–109. https://digitalcommons.law.yale.edu/yjil/vol22/iss1/2/.

Selling, Joseph A. 2016. *Reframing Catholic Theological Ethics.* Oxford: Oxford University Press. 10.1093/acprof:oso/9780198767121.001.0001.

Siegel, Shefa. 2013. "The Missing Ethics of Mining." *Ethics and International Affairs* 27, no. 1. https://www.ethicsandinternationalaffairs.org/2013/the-missing-ethics-of-mining-full-text.

Simmons, Ernest J. 1946. *Leo Tolstoy.* Boston: Little, Brown and Company.

Stahn, Carsten, and Jann K. Kleffner, eds. 2008. *Jus Post Bellum: Towards a Law of Transition from Conflict to Peace.* The Hague: TMC Asser Press. 10.1007/978-90-6704-719-7.

Steffen, Lloyd. 2012. *Ethics and Experience: Moral Theory from Just War to Abortion.* Lanham, MD: Rowman & Littlefield, 2012.

United Nations Human Rights Office of the High Commissioner. 2014. *Frequently Asked Questions About the Guiding Principles on Business and Human Rights.* Geneva: United Nations. https://www.ohchr.org/documents/publications/faq_principlesbussinesshr.pdf.

United Nations World Commission on Environment and Development. 1987. *Our Common Future: Report of the World Commission on Environment and Development.* Oslo: United Nations. https://sustainabledevelopment.un.org/content/documents/5987our-common-future.pdf.

Vocabulary.com. n.d. "Extraction." Accessed April20, 2021. https://www.vocabulary.com/dictionary/extraction.

Winright, Tobias. 1995. "The Perpetrator as Person: Theological Reflections on the Just-War Tradition and the Use of Force by Police." *Criminal Justice Ethics* 14, no. 2: 37–56. 10.1080/0731129X.1995.9991997.

Winright, Tobias. 2009. "The Liturgy as a Basis for Catholic Identity, Just War Theory, and the Presumption against War." In *Catholic Identity and the Laity*, edited by Tim Muldoon, 134–151. Maryknoll, NY: Orbis Books.

Winright, Tobias. 2016. "Your 'Just Peace' Reading List." *National Catholic Reporter*, December 21, 2016. https://www.ncronline.org/books/2017/08/your-just-peace-reading-list.

Winright, Tobias. 2017. "Peace on Earth, Peace with Earth: *Laudato Si'* and Integral Peacebuilding." In *All Creation Is Connected: Voices in Response to Pope Francis's Encyclical on Ecology*, edited by Daniel R. DiLeo, 195–211. Winona, MN: Anselm Academic.

Winright, Tobias. 2020. *Serve and Protect: Selected Essays on Just Policing*. Eugene, OR: Cascade Books.

Wirzba, Norman. 2012. "Reconciliation through Christ." In *Making Peace with the Land: God's Call to Reconcile with Creation*, edited by Fred Bahnson and Norman Wirzba, 61–82. Downers Grove, IL: InterVarsity.

Zenner, Christiana [published as Christiana Z. Peppard], and Andrea Vicini. 2015. "Introduction: On Just Sustainability and Its Challenges." In *Just Sustainability: Technology, Ecology, and Resource Extraction*, edited by Christiana Z. Peppard and Andrea Vicini, 1–10. Maryknoll, NY: Orbis Books. 10.1353/sce.2018.0018.

9 Integral ecology, just peace, and mining

Anna Floerke Scheid and Daniel P. Scheid

Pope Francis' encyclical *Laudato Si'* (2015) represents a vital expansion and development of Catholic social teaching on ecology. In particular, "integral ecology" provides a valuable moral framework that can help articulate norms for building a just peace in light of the modern extractive industry. Early Catholic teaching on our moral duties to safeguard the "environment" or "natural environment" traditionally employed a simple, utilitarian approach to nonhuman goods: they should be shared amongst humankind and used to promote human well-being. This initial approach has blossomed into a much more sophisticated theology of creation, and an analysis of the multilayered causes of ecological degradation, encapsulated by Francis as "integral ecology." Integral ecology proposes a broader holistic telos for human interaction with nonhuman plants, animals, ecosystems, and the planet as a whole (and indeed the entire universe). Moreover, integral ecology intersects with Catholic teaching on war and peace and provides the foundation for an ecological expansion of the idea of just peace, yielding an *ecological* just peace. This essay will proceed in three parts: first, an overview of the main principles of integral ecology; second, a description of the principles and practices that build a just peace; and third, a brief application of integral ecology and just peace to mining.

Integral ecology

The two central themes of *Laudato Si'* (Francis 2015) are the Gospel of Creation and integral ecology. The Gospel of Creation outlines the scriptural and theological roots for affirming the goodness of creation and the importance of the non-human world for the human understanding of God and salvation history. The Gospel of Creation provides a key set of theological themes such as seeing creation as an "order of love," affirming the Earth not just as nature but as creation, and perceiving the indwelling of the resurrected Christ in all creatures (§77, 75, 83). Integral ecology is a synthetic term that formulates an ethical application of the theological claims of the Gospel of Creation and its inextricable connection to social justice and Christian ethics. While all the principles of

DOI: 10.4324/9781003094272-9

Catholic social teaching are important for addressing a Catholic peacebuilding approach to mining, integral ecology speaks to the shift in moral vision required to address it adequately.

One of the key drivers of the ecological crisis for Pope Francis, and by extension at the nexus of mining and violence, is the technocratic paradigm. Humans have used technology to alter their immediate surroundings for millennia, just as humans have engaged in mining across civilizations for centuries. Yet technology has now become so powerful and seamless in our globalized cultures that it has led to a new way of seeing the world. The "technocratic paradigm" is a one-dimensional approach to the world that encourages people to disregard and overcome natural limits. For all the wonderful gifts it affords, technology, Francis reminds us, is not neutral. Technology creates a physical and mental framework that conditions lifestyles and social relationships. The technocratic paradigm encourages people to see the world as formless, without any inherent guidelines as to how it is to be used. Thus, this way of framing envisions the entire world as open to human mastery, control, and manipulation (§106–110).

By contrast, the principle of integral ecology configures the world differently, emphasizing reality as holistic, interconnected, and fundamentally relational. Integral ecology stems from the Gospel of Creation and from the theological conviction that the universe is created by a God who is Trinity and essentially relationship (§240). Yet like all principles of Catholic social teaching, integral ecology also belongs to the natural law and is accessible to people of various backgrounds who do not share this same Catholic and Christian understanding of the Creator. Here we will emphasize three aspects of integral ecology.

First, Francis introduces integral ecology as a holistic moral framework. If the central tendency of the technocratic paradigm is to narrow our moral vision, then integral ecology insists on expanding it. We must return to a "broader vision of reality," otherwise all our vast and various storehouses of information and insight can themselves become a "form of ignorance" (§139). Integral ecology begins with and privileges the whole, rather than starting with any particular part, because it posits that the whole and the relationships that form within it are inextricable from understanding the meaning and purpose of each part. Therefore, one cannot look at the world, or more specifically, the question of the value of mining, from the perspective of humanity alone (or worse, from one subset of humanity and its endless desires) without also looking at the whole that enables human beings to exist and to live peaceably and harmoniously with each other and the rest of creation.

Isolating the economic interests of one country over another, or of one segment of the economy, such as mining, vis-à-vis another, may be useful to a degree and for a limited period of time, but only if the overriding context of the whole is elevated throughout and returned to repeatedly. The whole must be considered before any one part is considered in isolation. In a way,

Francis challenges modern tendencies to consider "nature" or "the environment" as something outside of human beings and human culture. The tendency to separate concerns about the environment from other human concerns, or from economic and political issues, is part of a reductive moral vision. There is only one integrated reality, and we must see humans and non-human creation alike as part of it.

Second, integral ecology posits this broader vision of reality as pervasively interconnected. We cannot entirely separate one part from another because the strands that connect us to each other are innumerable. Again, we can consider the interests of a particular group separate from the health of other groups, but only for a time. Because all creatures are interconnected, any decision or choice made in one area will affect others. We live in a hybrid world insofar as categories like "nature" and "society" are not strictly separate, but interpenetrate one another. Therefore, we cannot, for example, divorce human wellbeing from the wellbeing of the planet as a whole, or social inequities and the violence that can result from them from ecological pressures. "A true ecological approach *always* becomes a social approach; it must integrate questions of justice in debates on the environment, so as to hear *both the cry of the earth and the cry of the poor*" (§49). For Francis, the suffering of the poor is connected to the suffering of the planet, and both must be addressed simultaneously. This is one of the central moral claims of *Laudato Si'*.

Integral ecology, which fuses social and ecological justice, introduces salient moral norms like the danger of a "throwaway culture" which reduces the poor and the Earth itself to "rubbish" (§22). We have not adopted a circular approach to our economy in which waste products are reabsorbed into the production of necessary goods. A throwaway culture is the ethical counterpoint to integral ecology, and it signals the need to address the degradation of the Earth and the hardships of the poor simultaneously. Francis particularly highlights the cultural threats faced by indigenous peoples and calls them "the principal dialogue partners" for economic and political elites (§146). Francis's attention to indigenous communities is an intensification of the links between social and ecological injustice, and resultant violence. These links are especially important when assessing the ethical ramifications of mining, since indigenous peoples have long been marginalized from their lands to make room for economic development that does not benefit them. Integral ecology underscores the interrelatedness, and even interdependence, of all Earth's creatures and systems. Likewise, interconnectedness describes issues of poverty, violence, mining, and ecological degradation. We cannot neatly or simply separate economy from culture, social justice from ecological justice, or nation-states from multinational corporations.

Third, integral ecology is relational and is meant to incorporate ecology into the everyday rhythms of human life. For this reason, Francis connects ecological concerns like climate change and biodiversity to the full range of human experience and makes them personal. Along with threatened ecosystems and creatures, integral ecology recognizes the threats faced by various cultural heritages. Francis frames integral ecology in this holistic and relational sense by pointing out that it inculcates a sense of belonging: "There is also a need to protect those common areas, visual landmarks and urban landscapes which increase our sense of belonging, of rootedness, of 'feeling at home' within a city which includes us and brings us together. It is important that the different parts of a city be well integrated and that those who live there have a sense of the whole, rather than being confined to one neighborhood and failing to see the larger city as space which they share with others" (§151). Here we see all three aspects of integral ecology: we must begin with the sense of the whole and what brings us together; we must see different parts of the city as connected; and finally this brings to us a deeper sense of our relationships to each other and the world around us, with the goal of helping us feel rooted and "at home" in the world. The integral ecology of everyday life feeds into other principles of Catholic social teaching, such as the common good and solidarity with future generations.

Integral ecology represents an expanded moral vision, the kind necessary to break through the restricting and stultifying lens of the technocratic paradigm. Indeed, Francis points to an ultimate dimension of integral ecology that stretches to the limit of time and space. Creation, recall, is a larger category with a broader meaning than nature, "for it has to do with God's loving plan in which every creature has its own value and significance" (§76). Francis links the Creator's love of every creature to eternity. The church's tradition teaches that all of creation is ordered by God's love, encouraging us to move from the love of the world around us toward the source of this world and the Creator's love. Francis also stretches our sense of time by including the eschatological horizon, referring to the ultimate destiny of all creatures transformed by God (§243).

Integral ecology, then, impels us to consider the consequences of our actions presently and how they may unfold over the centuries. Solidarity with future generations is a sensible virtue in light of the manifold connections we have across time and space. Francis encourages us to see the common good not just of particular nation-states or of the human family but of the planet and the cosmos as well. For many, such a moral vision will prove daunting; bearing in mind our shared eschatological future or the full cosmic import of our choices will not fully or even perhaps partially determine concrete practical political solutions toward peacemaking and social justice. But integral ecology does point to a much wider moral vision, suggesting that real ethical norms and choices ought to be rooted in this broader cosmic worldview if they are to be aligned with the Catholic understanding of a good creation stemming from a loving Creator.

Rooted in the Gospel of Creation, the ethic of an integral ecology unites human and planetary well-being into a common vision. It provides the context for other principles of Catholic social teaching, such as the common good, solidarity, and integral human development, as they apply to issues of mining, violence, and peacemaking. By stressing the interrelatedness of humans and nonhumans, and of the suffering of the Earth with the suffering of the poor, integral ecology provides a foundation for discussing ecological debt and the obligations that the Global North has to the Global South when evaluating the relationships at stake in the mining industry and in light of the historical role that mining operations have had in creating current economic and financial imbalances between the Global North and South. Integral ecology is consonant with what scientists tells us is a rapidly changing world, both ecologically and politically. Integral ecology would encourage a comprehensive and scientifically sound plan for properly balancing mining's benefits and harms in a warming world that is accelerating conflict. Now we turn to an area of longstanding Catholic teaching that converges well with Francis' depiction of integral ecology, namely the Christian tradition of "just peace."

An ecological just peace

The notion of "just peace" has been developed in Judeo-Christian tradition over the course of millennia. In the New Testament, a just peace is the aim of Jesus' efforts to usher in the Reign of God. In God's reign "the poor [will] no longer be poor, the hungry [will] be satisfied, and the oppressed [will] no longer be miserable" (Nolan 1976, 58). To live out the Reign of God is to live a just peace, sharing our possessions, serving the "least" of God's people, and loving our neighbors as we love ourselves. This just peace envisioned by Jesus as he preached the Reign of God has roots in the biblical notion of *shalom,* the deep peace articulated in the Hebrew Bible, founded on right, or just, relationships between humankind and God, amongst humans themselves, and between human beings and the rest of God's creation. This last component is mostly latent and undeveloped in the church's tradition, but it bespeaks the potential for the tradition of just peace to dovetail with Francis's "integral ecology." Both integral ecology and just peace are rooted in foundational Christian theological principles of the goodness of creation and the central message of the Kingdom of God, and so they can combine to contribute to peacebuilding in the midst of violence resultant from unjust mining industry practices.

Pope St. Paul VI and Martin Luther King, Jr., contemporaries, were both important figures in bringing the biblical notion of just peace into the political arena in the mid-twentieth century. Their legacy of understanding peace as constitutive of justice remains critical today. Paul VI (1967, §76) famously insisted that "peace is not simply the absence of warfare, based on a precarious balance of power; it is fashioned by efforts directed day after day toward the establishment of the ordered universe willed by God, with a more perfect form of justice among men."

In a similar way, King (1963) responded to those who accused him of fomenting violence by distinguishing between positive and negative peace. The American Civil Rights Movement, with its nonviolent methods of protest and civil disobedience, did not *cause* tension or violence, but merely surfaced those tensions already present, and absorbed oppressive people's and systems' violent outbursts. While a lack of outright bloodshed may, then, signify a negative peace, King argued that a positive peace is indicated by the presence of justice.

More recently, Christian scholars have refined the concept of just peace considerably (Dennis 2018; Cahill 2019; McCarthy 2020), further arguing that a just peace aims at (a) preventing violence from breaking out in the first place, and (b) defining and refining principles and (c) practices of just peace. The question of mining and peacebuilding suggests the need for a new consideration of the idea of just peace, one that explicitly includes ecological sustainability and restorative justice for the planet. An *ecological just peace* would insist on the ultimate goal of developing *only* renewable sources of energy, and on approaches to human dignity and community, work and economics, that begin and accord with the vision of an ecological just peace. The eschatological horizon of integral ecology expands the reasonable timeframe for considering economic decisions, so that while some mining is essential for renewable energy, the consequences for this activity, which stretch out into centuries, must be factored in. The burden of proof for justifying the ecological harms of mining must be made against this ethical backdrop, making short term financial gains increasingly difficult to warrant. Accordingly, in the next section, we explore emerging ideas about just peace, and their salience for establishing an ecological just peace.

Prevention

A just peace establishes mechanisms for preventing violence from breaking out in the first place, and interrupts cycles of violence. Just peacemaking theory (JPT) consists of ten practices meant to stave off war or violence (Stassen 2008). One resonates deeply with integral ecology: foster just and sustainable economic development. This practice aims to prevent violence and maintain a just peace through locally controlled economic development. "Commitment to the sustained well-being of human beings everywhere, and of their local, regional, national, and global communities is vital to justice and peace and to care of the earth…in sum, justice and peace are closely bound together with sustainable development" (Bronkema, Lumsdaine, and Payne 2008, 134–35). JPT authors suggest that the most successful sustainable development efforts happen at the microlevel with the help of non-governmental organizations. This presents an obvious challenge to any mining industry model that privileges large corporations as potential partners with national governments in sustainable development to the exclusion of smaller and more locally sourced enterprises.

The prevalence of arrangements in which national governments contract with multinationals to extract resources in discrete regions with little input from local leaders leads scholar Philippe Le Billon (2008, 346) to argue that "studies of resource-related conflicts need to broaden their analysis beyond spaces of resource exploitation to include the interrelationship between spaces of production, consumption, representation, and governance." In other words, fostering an ecological just peace requires attention not only to land, but to the nexus of land, culture, politics, and government. Le Billon's remark echoes the inter-relationality so critical to Francis's notion of integral ecology described above. Similarly, Peter Knox (2015) points to the challenges of conducting mining justly by highlighting the lingering effects of colonialism and apartheid in South Africa. The South African regime heavily subsidized white people, for example by granting mineral rights to landowners, but only allowed thirteen percent of land to be owned by black South Africans (121). Randy Odchigue (2015) notes a similar problem in the Philippines, where mining not only threatens biodiversity hotspots but also disenfranchises indigenous peoples, who are among the poorest, of their identity and culture. "The mining industry of the Philippines has yet to show an example of a community that justly and sustainably thrives after its resources are extracted and exported to other countries in the world. This example reveals how degradation of the environment is directly proportional with the exploitation of the poor" (171–72). In this book, Elias O. Opongo's chapter on mining governance and the challenges of equitable distribution of mining profits argues for including local populations in policy-setting processes in order to achieve development that truly improves communities' standards of living.

At the same time, it is important to be mindful that this JPT practice of fostering just and sustainable economic development was articulated prior to Francis's discussion of integral ecology. Along with other pleas for "sustainable development," it has been critiqued as possibly damaging to the environment. For example, the co-author of this essay has drawn on the work of Leonardo Boff to argue that peacebuilding requires not so much new development as virtuous practices of "dignified subsistence" (D. Scheid 2012). Furthermore, the language of "development," especially when it arises from Western contexts, may simply be too embedded with violent colonial and neocolonial assumptions to be helpful for fostering local leadership in indigenous contexts. This just peacemaking principle, then, remains important for preventing violence, but it ought to be expanded further through dialogue with integral ecology, decolonial theory, and indigenous local leadership in discrete contexts.

In light of this discussion of sustainable development and just peace, then, several difficult questions must be considered if we want to promote an ecological just peace. Considering intensifying climate change, tailings dam collapses, acid mine drainage, and other forms of environmental harms, to what degree must any understanding of the concept of "just peace" require ecological sustainability? More specifically, can the mining industry develop in any way that is truly both just and sustainable? Modern mining processes seem to match Francis's critique of the technocratic paradigm in terms of the control and domination of the land, and concomitant efforts to dominate the poor. In this way, there is often an inherent violence on ecological systems and indigenous people within some mining systems that themselves help constitute part of the ecological crisis. Given the telos of integral ecology which seeks healthy long-term relationships amongst human beings, the natural world, and planetary systems, how might the mining industry modify its practices in order to contribute to the transition to a truly just and sustainable economic system? There are, for example, efforts at long-term containment of waste following mine closures and at the restoration of sites by modern mining companies (see, for example, ICMM n.d.). An ecological just peace supports these efforts but also critically evaluates them by setting the goal of fully sustainable and renewable practices, with an ethical horizon of millennia rather than decades. Mining must move beyond mere legal compliance or responding to immediate social concerns and set for itself this broader and more substantive standard for action.

Principles

Catholic scholars have begun to name principles that undergird a just peace. Maryann Cusimano Love (2010, 56–57) enumerates several principles which she calls "norms of just peace" that function in ways akin to and inspired by the criteria of the Christian just war tradition. Love suggests that all Catholic peacebuilding ought to include: just cause, that is defending human dignity and the common good; right intention toward a positive peace; participatory processes, or making room for all stakeholders to contribute to peacebuilding efforts; restoration of social relationships; reconciliation, understood as addressing and acknowledging the harms of violence; restoration of infrastructure; and sustainability, developing structures that can help peace endure over time. Combining Love's principles and Francis's integral ecology, we suggest two foundational principles for an ecological just peace: respect and restoration.

Respect

Drawing on Love's principles of just cause, participatory process, and sustainability, and Francis's (2015, §22) critique of the technocratic paradigm and a throwaway culture which dismisses the dignity of others on behalf of convenience, we argue that just peace is built on a foundation of *respect*. Its ground is the innate dignity that accrues to every creature by dint of its relationship to the Creator, while there also remains a particular elevated dignity for human beings (Francis 2015, §65, 69). This respect must be built up through both horizontal and vertical dimensions of social life, reaching not only large governmental and multinational organizations, but also impacting communities and personal daily decisions. Ideally integral ecology and an ecological just peace become a kind of culture, and thus an expectation, that is participatory and intergenerational.

An ecological just peace rooted in respect rejects piecemeal responses to discrete environmental crises, and to communities' concerns about such crises. "Ecological culture cannot be reduced to a series of urgent and partial responses to the immediate problems of pollution, environmental decay and the depletion of natural resources" (§111). In the same way, an ecological just peace cannot simply react to discrete outbreaks of violence that result from discrete extractive practices. Rather, an ecological just peace must imbue entire cultures: "There needs to be a distinctive way of looking at things, a way of thinking, policies, an educational programme, a lifestyle and a spirituality which together generate resistance to the assault of the technocratic paradigm" (§111). A peace that is built with justice and is truly ecological is committed to building a society that *respects* the common good, or the well-being of the community itself; the dignity of each human being; and the well-being of the planet, its resources, creatures, and ecosystems, or the "splendid universal communion" of all creatures (§220). This final element returns to Francis's holistic and cosmic vision, which spreads far wider than the technocratic paradigm and its emphasis on piecemeal reactions as extractive-related violence and other crises explode.

Restoration

Second, drawing on Love's (2010) principles of right intention, reconciliation, restoration, and, again, sustainability, and Francis's advocacy for ecological education, we note that a just peace is ordered toward *restoration*. This restoration depends upon the foundational principle of respect. Moreover, restoration is comprehensive of shalom, meaning that it applies to relationships among human beings; between individual human beings and God, as well as human communities and God; and finally, between human beings and creation. Francis (2015, §210) calls for an ecological education that honors and restores harmony among the various relationships among human beings and the planet: "[An ecological education] seeks also to restore the various levels of ecological equilibrium, establishing harmony within ourselves, with others, with nature and other living creatures, and with God." Harms to individual persons, to human communities, to human infrastructure, and to ecosystems must all be acknowledged and repaired as best as possible for a sustainable and ecological just peace to take root and flourish. Many communities have cleaned rivers, restored forests, beautified landscapes, and enhanced urban settings with creative buildings; "these achievements do not solve global problems, but they do show that men and women are still capable of intervening positively. For all our limitations, gestures of generosity, solidarity, and care cannot but well up within us, since we were made for love" (§58).

Restoration also includes holding accountable corporations that share responsibility for violence related to mining. Some human rights lawyers suggest "a hierarchy of three forms of corporate complicity in order to distinguish the different ways ... companies might be held accountable for violations from which they ultimately derive an advantage" (Hilary 2017, 81). These are silent complicity—when a corporation "fails to speak out" about violence affecting the areas in which they operate; beneficial complicity—when a corporation "is the beneficiary of" violence done by state or militia forces; and finally direct complicity—when a corporation "provides assistance" to other people or groups knowing they are likely to commit violence. Understanding various forms of complicity in violence is critical to grasping how corporations are involved in harm to people and communities, developing mechanisms for holding them accountable, and envisioning how they can work to make amends.

Practices

It is helpful to identify practices that promote the principles of respect and restoration, and thus an ecological just peace. Our discussion here is far from comprehensive, but it provides a trellis up which the vines of ecological justice may grow and flower. We discuss nonviolent direct action, trust-building, imaginative thinking, and indigenous peacebuilding. We close this section with a recent example of a community struggling to enact an ecological just peace.

Nonviolent direct action

Peace studies scholars from multiple disciplines have been reminding us of the relative effectiveness of nonviolent direct action (Chenoweth and Stephan 2011; Nepstad 2011). Nonviolent direct action is actually an umbrella term for a number of resistance practices that address injustice without resort to violence, including marches, rallies, protests, boycotts, and civil disobedience. Nonviolent direct action as a means of surfacing and confronting serious injustices is more promising than armed action because it leads to comparatively fewer harms (particularly in the form of death, serious injury, torture, etc.). As King (1963) explained, these actions do not create discord and violence but expose the injustice and implicit violence that was present throughout and that those who benefit from or are not subject to structural violence often prefer to ignore (see Ethicists Without Borders 2019).

Nonviolent direct action is already a major feature of resistance to ecological injustice and mining-related violence around the globe. Arce and Moran (2020) detail the program of nonviolent direct action motivated by fears of ecological damage at the Tía María copper mines in Peru. There, nonviolent direct action helped push both the government and mining project developers to include the local community more robustly in negotiation processes (i.e., respect), ultimately leading the government to delay the mine's opening to 2024. Efforts to ensure a participatory democratic process around the copper mine continue (81–85), and the chapter in this book by José Bayardo Chata Pacoricona details how Derechos Humanos y Medio Ambiente–Puno focuses on legal action in other cases in Peru as a form of nonviolent direct action to avoid having protest situations reach a point where violence can occur.

Trust-building

An ecological just peace depends on building trust among stakeholders. Lisa Sowle Cahill (2019) argues that trust-building involves holding those responsible for harm accountable (i.e., restoration), as well as shared "activities together around shared goals." Such shared activities and shared goals are critical for the "practical validation of social trust" amongst people and communities in conflict (340). Similarly, Francis (2015, §213) calls for everyone to be involved in ecological education, from individuals and families, to churches, schools, businesses, and governments. Each has a role to play in creating a "a culture of shared life." Laws, however valuable and important, can only be effective in the long term if there are a sufficient number of people who are prepared to accept them (§211). Actions of beautifying, protecting, and restoring public places that belong to everyone (i.e., respect and restoration) are important because through them "relationships develop or are recovered and a new social fabric emerges. Thus, a community can break out of the indifference induced by consumerism. These actions cultivate a shared identity, with a story which can be remembered and handed on" (§232). Practices that build trust can contribute to a culture of shared life and to people who become ready to adopt new laws and regulations.

Imaginative thinking

Many scholars point to the importance of nourishing individual and collective human imagination in order to promote just peace. Preeminent peace studies scholar John Paul Lederach (2005, 172) urges:

> [I]f we are to survive as a global community, we must understand the imperative nature of giving birth and space to the moral imagination.... We must face the fact that much of our current system for responding to deadly local and international conflict is incapable of overcoming cycles of violent patterns precisely because our imagination has been corralled and shackled by the very parameters and sources that create and perpetrate violence.

Both respect and restoration depend on the capacity to think imaginatively in order to break out of entrenched cycles of violence and vengeance. In a similar vein, Robert Schreiter (2010, 221) notes that "the capacity to imagine peace, that is, to think differently about the conflict situation in order to come to new possibilities that might end the conflict, is now being recognized as one of the most important qualities of a peacebuilder."

This emphasis on imaginative thinking suggests that practices for building an ecological just peace will include, in Cahill's (2019, 332) words, "song, story, art, and ritual" that contribute to the "transformation of imaginations and worldviews so that a different reality is grasped as truly possible." How might artists, poets, musicians, and playwrights help restore and enrich our understanding, for example, not only of the efficiency and sustainability of renewable energy, but also the sheer beauty of it? Solar panels turn sunshine into our softly glowing baby monitor, so that the light that brings life to the planet also illuminates the doubts and fears of new parents and helps them feel secure. Turbines harness the wind rolling across the ocean to power an electric respirator, so that the invisible force of air itself breathes life into a critically ill patient.

Adjacent to these artistic ideas is the social imaginary of Catholicism itself. Schreiter (2010, 221) defines the social imaginary as "framed by certain assumptions about the world and certain rules of connection and communication…filled with certain values, images, and practices." How might Catholic imaginative thinking shape our capacity to envision and then build an ecological just peace? Here the Gospel of Creation is helpful. It posits Catholic imagination as fundamentally ecological, and provides ways of conceiving the nonhuman nature in all its variety as integral to human self-understanding and our relationship to the divine. Francis (2015, §96–98) imagines Jesus's relationship to the natural world, attentive to the beauty of the Earth and the loveliness of even the smallest creatures, living in harmony with all of nature. For Francis, Jesus embodies integral ecology, fusing a concern for justice and peace with a love for creation and the Creator from whom all things come.

Indigenous peacebuilding

Finally, an ecological just peace must include robust efforts to draw on values, beliefs, and peacebuilding practices indigenous to the locales in which the efforts occur. For issues related to mining violence, this will mean building and recognizing leadership amongst grassroots peacemakers who understand well the indigenous culture and its conflict transformation practices. Such practices often include reverence for or relationship with elements of the natural world (i.e., respect) and explicitly promote restoration. In Liberia, the palaver "is a ritualized practice of dialogue and communication, which enables the community to come to consensus around ethical norms and the resolution of conflicts" (A. Scheid 2012, 35).

This peacebuilding ritual often takes place under the community's "palaver tree," sacred ground for people to gather, discuss their differences, and restore justice. Liberians turned to the palaver in their struggle to embrace a just peace in the aftermath of a war that decimated communities through the use of child soldiers (Cahill 2019, 333).

Similarly, Rwandans have been using traditional Gacaca Courts to hold accountable those who participated in the 1994 genocide. The word "gacaca" simply means "a soft green bed of grass," and it is in such natural locations that the community gathers to ritually discuss and resolve conflicts. Gacaca Courts emphasize restorative justice and interpersonal and social reconciliation. Over two million Rwandans have been tried in Gacaca Courts since the genocide. Likewise, South Africa wrote the principle of ubuntu into its first democratic constitution in 1994, and promoted it through the Truth and Reconciliation Commission in the aftermath of apartheid. Ubuntu is a Southern African humanistic philosophy that emphasizes the social nature of the person, and the interdependence of the human community based on respect for our shared humanity. It translates roughly as "a person is a person through other persons," and it continues to be a vital force in South Africa for cultivating a culture that will value justice and peace as part of ongoing healing from apartheid.

El Salvador: a recent example of ecological just peacemaking

A recent example of a community struggling to enact an ecological just peace that honors the moral vision of integral ecology is the effort to ban metallic mining in El Salvador that is detailed in the chapter of this book by Andrés McKinley. In 2017, legislation passed that banned metallic mining following years of community efforts at organizing and raising awareness. Like many Central American countries, El Salvador has experienced centuries of mining and extractive operations conducted by foreign powers. Despite being among the poorest and most densely populated countries in the Western Hemisphere, and after suffering decades of violence and political repression, the people of El Salvador organized to ban metallic mining for a variety of concerns but primarily because of the effects on the quantity, quality, and access to safe water. McKinley describes how El Salvador is in the midst of a water crisis, and how climate change will only intensify the causes in the coming decades. Metallic mining requires huge stores of freshwater, and mining contaminates water in a variety of ways, most problematically acid mine drainage.

As McKinley observes, opposition to metallic mining in El Salvador evolved over seventeen years, and the many stages of this effort incorporated various elements of an ecological just peace. It began in 2000 with the growing presence of transnational mining companies. Local communities were already poised to respond because of their social bonds forged through years of resistance to companies prospecting for gold and silver and through their shared heartbreak during the brutal civil war. From the beginning, the Catholic Church was involved in galvanizing local communities to employ nonviolent means to resist foreign mining interests. The Jesuit-led University of Central America (UCA) conducted research that provided an extensive foundation of information on the long-term effects of the mining operations. Once a legal bill, developed in partnership with environmental experts at UCA, was proposed, the Catholic Church helped lead a massive street march in 2017 to persuade legislative leaders to pass it. When it passed, El Salvador became the first country to ban this form of mining.

El Salvador's resistance demonstrates key features of an ecological just peace: forging diverse coalitions through relationship building over a period of many years; using religiously inspired values and an expansive moral imagination to generate a public consensus; disseminating information and countering propaganda campaigns by foreign companies; and protesting nonviolently in the streets to garner the attention of elected leaders. The seventeen-year-long struggle provides a powerful example of how nonviolent direct action, coalition building, and the Catholic imagination motivated by a moral vision analogous to integral ecology, can promote and further an ecological just peace.

Ecological just peace and mining

The mining industry is diverse and multifaceted, and each mining enterprise has its own particular set of concerns, such as the target material of the extraction and the method used to extract it, the land it operates on, the historical and political context of the people who live there, the ecosystems of the region, the expectations of the particular extractive companies, etc. This essay has provided a general ethical framework that can situate particular extractive operations in a comprehensive Catholic moral vision, yet we can still draw some brief conclusions from this expansive perspective.

Any process of mining that is not rooted in a broad framework—like integral ecology—with a deep moral horizon will likely thwart efforts toward an ecological just peace. As Francis contends, modernity is pervaded by a technocratic paradigm, and this includes our appetite for extractives. Absent resistance, extractives will contribute to an intensification of a throwaway culture with its attendant waste, violence, and needless death. While the paradigm of integral ecology stresses a holistic, interconnected, and fundamentally relational framework of human-Earth relationship, the absence of an ecological just peace will lead to greater violence against ecosystems, the poor, and future generations. Without the model of integral ecology, land, ecosystems, and often even people are treated as parts that can be utilized and profited from, consumed or used to varying degrees, and then disposed of. Indigenous populations that stand in the way of development can be easily bypassed unless their voice is intentionally privileged and welcomed. In turn, environmental degradation and resource scarcity will become further causes of injustice, setting up cycles of ecological degradation, social conflict, and violence.

An ecological just peace draws on the principles of respect and restoration that undergird peacebuilding practices of nonviolent direct action, trust-building, imaginative thinking, and indigenous practices of reconciliation and restoration. This ethic involves appreciating the relationship of all creatures and ecosystems to their Creator, before measuring the economic utility of extraction; assessing and cataloguing the interrelatedness of an ecosystem and its people and predicting possible implications; valuing the long-term economic, psychological, emotional, and social benefits that come from a healthy and sustainable ecosystem; and welcoming input from local peoples and making them partners in decision making. The moral vision of integral ecology yields a framework for an ecological just peace that can help mitigate violent conflict and other forms of injustice in contexts of mining.

References

Arce, Moisés, and Riley Moran. 2020. "Extractive Conflicts in the Developing World." *Journal of International Affairs* 73, no. 2: 77–94.

Bronkema, David, David Lumsdaine, and Rodger A. Payne. 2008. "Foster Just and Sustainable Economic Development." In *Just Peacemaking: The New Paradigm for the Ethics of Peace and War*, edited by Glen Stassen, 132–152. Cleveland, OH: The Pilgrim Press.

Cahill, Lisa Sowle. 2019. *Blessed Are the Peacemakers: Pacifism, Just War, and Peacebuilding*. Minneapolis: Fortress Press.

Chenoweth, Erica, and Maria J. Stephan. 2011. *Why Civil Resistance Works: The Strategic Logic of Nonviolent Conflict*. New York: Columbia University Press.

Dennis, Marie. 2018. *Choosing Peace: The Catholic Church Returns to Gospel Nonviolence*. Maryknoll, NY: Orbis.

Ethicists Without Borders. 2019. "Christian Ethics, Climate Emergency, and Nonviolent Direct Action." June 9, 2019. https://dailytheology.org/2019/06/09/statement-of-ethicists-without-borders-christian-ethics-climate-emergency-and-nonviolent-direct-action/.

Francis. 2015. *Laudato Si'*. http://www.vatican.va/content/francesco/en/encyclicals/documents/papa-francesco_20150524_enciclica-laudato-si.html.

Hilary, John. 2017. *The Poverty of Capitalism: Economic Meltdown and the Struggle for What Comes Next*. London: Pluto Press.

ICMM. n.d. "Integrated Mine Closure—Good Practice Guide." Accessed May11, 2021. https://guidance.miningwithprinciples.com/integrated-mine-closure-good-practice-guide/.

King, Martin Luther, Jr. 1963. *Letter from a Birmingham Jail*. http://www.africa.upenn.edu/Articles_Gen/Letter_Birmingham.html.

Knox, Peter. 2015. "Sustainable Mining in South Africa: A Concept in Search of a Theory." In *Just Sustainability: Ecology, Technology, and Resource Extraction*, edited byChristiana Z. Peppard and Andrea Vicini, 117–130. Maryknoll, NY: Orbis.

Le Billon, Philippe. 2008. "Diamond Wars? Conflict Diamonds and Geographies of Resource Wars." *Annals of the Association of American Geographers* 98, no. 2: 345–372.

Lederach, John Paul. 2005. *Moral Imagination: The Art and Soul of Building Peace.* New York: Oxford University Press.

Love, Maryann Cusimano. 2010. "What Kind of Peace Do we Seek? Emerging Norms of Peacebuilding in Key Political Institutions." In *Peacebuilding: Catholic Theology, Ethics, and Praxis*, edited by Robert J. Schreiter, R. Scott Appleby, and Gerard F. Powers, 56–91. Maryknoll, NY: Orbis.

McCarthy, Eli, ed. 2020. *A Just Peace Ethic Primer: Building Sustainable Peace and Breaking Cycles of Violence.* Washingon, DC: Georgetown University Press.

Nepstad, Sharon Erikson. 2011. *Nonviolent Revolutions: Civil Resistance in the Late Twentieth Century.* New York: Oxford University Press.

Nolan, Albert. 1976. *Jesus Before Christianity.* New York: Orbis Books.

Odchigue, Randy J.C. 2015. "The Ecclesial Contribution to Sustainable Communities." In *Just Sustainability: Ecology, Technology, and Resource Extraction*, edited by Christiana Z. Peppard and Andrea Vicini, 171–181. Maryknoll, NY: Orbis.

Paul VI. 1967. *Populorum Progressio.* http://www.vatican.va/content/paul-vi/en/encyclicals/documents/hf_p-vi_enc_26031967_populorum.html.

Scheid, Anna Floerke. 2012. "Interpersonal and Social Reconciliation: Finding Congruence in African Theological Anthropology." *Horizons* 39, no. 1 (Spring): 27–49.

Scheid, Daniel P. 2012. "Just Peacemaking Theory and the Promotion of Dignified Subsistence." In *Violence, Transformation, and the Sacred: They Shall Be Called Children of God* (College Theology Society Annual Volume 2011), edited by Margaret R. Pfeil and Tobias L. Winright, 175–189. Maryknoll, NY: Orbis.

Schreiter, Robert J. 2010. "The Catholic Social Imaginary and Peacebuilding: Ritual, Sacrament, and Spirituality." In *Peacebuilding: Catholic Theology, Ethics, and Praxis*, edited byRobert J. Schreiter, R. Scott Appleby, and Gerard F. Powers, 221–239. Maryknoll, NY: Orbis.

Stassen, Glen, ed. 2008. *Just Peacemaking: The New Paradigm for the Ethics of Peace and War.* Cleveland, OH: The Pilgrim Press.

10 Mining, Catholic social teaching, and international human rights

Douglass Cassel

Mining worldwide generates a diverse range of risks and harms to human rights, the environment, and peace. These arise from extraction, transport, waste, and mining revenues, and vary depending on what is mined, where, how, and the end uses. Mining may also yield offsetting benefits in employment, economic growth, and valuable end uses such as medical equipment and personal computers. Catholic social teaching and secular international human rights norms on mitigating harms and enhancing benefits have much in common and are generally compatible. Both focus on the impacts of mining on human dignity, the environment, and peace. However, they also diverge in fundamental ways.

As epitomized in *Laudato Si'*, Catholic social teaching recognizes fundamental human rights and the need for international human rights laws and institutions (Francis 2015, §53, 93, 157, 173, 175). But the concept of human rights in Catholic social teaching is much broader than in secular law. The church insists on an integral, holistic, communitarian, spiritually grounded vision of human development and peace. Concern for our planet unites with compassion for our sisters and brothers, especially the poor, in an inseparable whole. These lofty goals call for nothing less than a cultural "revolution" (§114). *Laudato Si'* teaches that our societies, ecology, and personal well-being are all maimed by market-driven materialism, consumerism, overblown individualism, self-centered pursuit of immediate gratification, anthropocentrism, and a false faith that endless technological progress will forever yield unlimited economic growth. It sounds an urgent call that our earthly home is fast reaching the limits of its capacity to tolerate our cultural and material excesses.

Secular international human rights norms share important values with the Catholic faith, notably respect for human dignity and security. They take broader stands in support of liberty and against discrimination. But they do not aim as high as the church. Widely agreed international norms are necessarily the product of consensus diplomacy among nations, whose diverse political and value systems are immersed in a market-driven global culture of technological progress and material growth. Norms adopted by international organizations reflect both what is common and what is diverse among their member states.

DOI: 10.4324/9781003094272-10

The United Nations encompasses democratic as well as repressive states, and religious as well as secular cultures. The result is that international norms, far from upsetting political and economic apple-carts, tend toward the lowest common denominator. Rather than reach for the heavens, they erect a floor. You need not love your neighbor, but you must not torture her.

Unlike Catholic social teaching, international human rights norms emphasize rights over responsibilities. Lacking an agreed, unifying theory, international human rights reduce to a menu of separate rights, instead of a cohesive, integral whole (see Glendon 1998). While necessary to avert suffering, they do not aspire to maximize the common good. They do not challenge the sins of their authors: materialism, consumerism, self-centeredness, and the myth of endless growth. They accept cultures, however misguided, and development models, however short-sighted. The price of gaining widespread acceptance by governments is settling for minimum, albeit critically important, rules of civility.

Mining is not exempt from these profound differences between Catholic social teaching and international human rights norms. If *Laudato Si'* is revolutionary, secular international norms for mining are merely regulatory. Coal mining is an example. *Laudato Si'* asserts the "use of highly polluting fossil fuels—especially coal...needs to be progressively replaced without delay" (Francis 2015, §165). By contrast, international norms seek to curb the coal industry's pollution, require it to respect local communities, avoid repressive violence, not fuel armed conflict, and move gradually to reduce carbon emissions. Still, at present, they seek only to rein in, not challenge the very existence of, the industry.

This difference between Catholic social teaching and international human rights norms is illustrated more broadly by their contrasting attitudes toward the end uses and profits of mining. The church considers the intrinsic value of the end use of what is mined, based on its contribution to integral human development. By contrast, international human rights norms on mining pay no attention to the intrinsic value of the end use. International norms adopted by governments, not by theologians, accept market valuations of mining based on monetary worth in consumerist societies.

Consider gold mines. Even when operated by the cleanest modern techniques, they pollute. Some level of environmental cost could be justified if the gold is to be used for, say, life-saving medical technology. In fact, the vast majority of gold mined nowadays (other than for bullion) is destined for jewelry (Sepanek 2012). The church would ask, is self-adornment worth the environmental costs? International human rights norms do consider the environmental impacts of gold mining. But they pay no attention to the intrinsic value of the end use. Their environmental regulations are the same regardless of whether the gold will be used to save lives or to adorn people. Much less do they condition the right to mine gold for a profit on the end use.

Catholic social teaching and international human rights norms also differ on the pope's injunction to mining executives to balance profits against a need to "reduce, reuse, recycle" (Francis 2019). There may be no present need to mine more gold for jewelry. There may be so much of the mineral already in jewelry stores and pawn shops that we do not need to dig for more. Gold-mining executives might choke on this conclusion, and to suspend gold mining for jewelry would also affect the gainful employment of miners, cutters, and merchants. Still, on a holistic view, Catholic social teaching arguably calls on us, before we mine more, to sell what already gleams through countless shop windows around the globe. International human rights norms do not go so far. So long as reasonable environmental and human rights safeguards are followed, these secular norms do not object to a mining company digging for gold we do not need in order to sell it for a profit. If a company can make money by mining, that's fine. And if a country's government, aiming to expand its economy, markets itself as the gold mining capital of a world already up to its neck in baubles, that's fine, too.

Catholic social teaching, then, is vastly more ambitious than international human rights norms. Even so, its lofty aspirations may have much to learn from the specificity and practicality of evolving international human rights norms for business, in general, and mining, in particular. The United Nations *Guiding Principles on Business and Human Rights* reiterate the legal duties of states to protect human rights, including from business; confirm the social responsibility of business to respect human rights; and make clear the duty of states and responsibility of business to facilitate remedies when business activities adversely impact human rights (UN Office of the High Commissioner for Human Rights 2011). The *Guiding Principles* are now increasingly supplemented and implemented by other global and regional norms, national laws, and, as Ray Offenheiser outlines in his chapter in this volume, by industry codes and company policies and procedures. Soon after their adoption, the church wisely endorsed the *Guiding Principles* (Tomasi 2014). Much remains to be done, but the trend lines of secular norms are moving toward moderating the worst social and environmental excesses of the mining industry. Yet it is unrealistic to expect secular norms, however helpful, ever to embrace the revolution in values, culture, worldview, and models of development advocated by *Laudato Si'*.

The following sections address, first, the diversity of mining industries; second, the impacts of mining on human rights, the environment, and peace; third, teaching from *Laudato Si'* pertinent to mining issues; fourth, international human rights norms; fifth, international norms on peace; and finally, how Catholic teaching and international norms might be combined to better serve peace and the common good.

Mining diversity

Not all mining is the same; neither Catholic social teaching nor international norms can adopt a one-size-fits-all approach. Analysis of costs and benefits must be disaggregated, and should consider the mining technologies used for different substances and varied end uses, ranging from frivolous to useful to critically important.

There is geographical diversity and endless variation in what is mined and end uses. China is by far the globe's champion miner. Not only is it the world's largest producer of coal and gold, China also has "great prospecting potential" for 22 other "major minerals," including iron ore, manganese, chromite, copper, lead, zinc, bauxite, tungsten, tin, molybdenum, antimony, nickel, silver, lithium, pyrites, sulphurite, phosphate rock, potash, magnesite, fluorite, boron and barite (Wu and Li 2018). In addition, China is the world's leading producer of "rare earth minerals" which are unknown to most of us, but are strategically or commercially important. Leading examples include lanthanum (used in hybrid vehicles and smart phones), cerium (used in catalytic converters and LED lighting), and terbium (used for solid state electronics, fuel cells, and naval sonar).

Latin American mining portfolios are likewise large and varied. Mexico, the world's largest miner of silver, is also an important producer of copper, coal, gold, zinc, and uranium, and a major producer of twelve minerals, three of which are important for U.S. demand: fluorspar (used in aluminum, gasoline, insulating foams, refrigerants, steel, and uranium fuel), graphite (for lubricants, electrodes, batteries, and solar panels), and strontium (for fireworks, flares, and glow-in-the-dark paints, among others) (UITA 2019b). Mining mixes differ elsewhere in Latin America. Chile leads the world in mining copper and is second in lithium (UITA 2021). Peru is Latin America's largest miner of gold and second largest of copper and silver (UITA 2020). Colombia leads the region in coal mining and has "great opportunities" for gold, copper, and nickel (Griffin and Acosta 2020). Important mining sectors in Bolivia include zinc, lead, tin, gold, silver, copper, and tungsten (UITA 2019a).

The foregoing merely gives a flavor of the diversity in what is mined, where, and for what end uses. The goal is not to present an exhaustive global portrait, but to illustrate differences that can affect how both Catholic social teaching and international norms assess mining industries.

Mining's global and local impacts

The social and environmental impacts of mining are both global (especially in end uses and through global warming) and local (the impacts on human rights, the environment, and peace in the community, country, and region where the mining takes place).

The most extensive global impacts at present arguably come from burning coal to fuel electric power plants and heat homes—a major source of carbon emissions and a major contributor to climate change. With substitute fuels increasingly available and competitive, thermal coal use globally appears to have peaked in 2013. Since then, use has fallen steeply in Europe and North America, and is projected to decline even further by 2025. However, in China, which already produces and consumes nearly half of all thermal coal worldwide, demand in 2025 is projected to be about the same as in 2019. Meanwhile, thermal coal use in India, and also in Southeast Asia, each of whose coal usage is already more than the United States and Europe combined, is projected to keep growing (International Energy Agency 2020). Other top revenue generators from mining—coking coal, iron ore, and copper—likewise contribute significantly to climate change (Greenpeace Australia Pacific 2017; Pooler 2019; Brown 2016).

Nuclear power is touted by some as a clean substitute for electricity generation by fossil fuels. But wholly aside from the significant risks of radioactive pollution from uranium mining, and the safety and environmental risks of operating nuclear power plants, global challenges of uranium mining include disposing of nuclear waste and ensuring that nuclear materials are used only for peaceful purposes, rather than for nuclear arms proliferation or for terrorism.[1]

In the communities and countries (and sometimes the regions) where mining takes place, the potential adverse impacts of mining on human rights, environment, and peace are many and varied. They may include discharge of toxic substances into air, water, and soil, harming not only human health, but the continued ability of local residents to farm, fish, earn a livelihood, and sustain their families. Vibrations and noise disturb tranquility and may crack walls and roofs of housing. Indigenous lands, sometimes including sacred places, are often encroached upon without free, prior, and informed consent. There may be little or no community consultation about mining projects. Importation of mineworkers may lead to prostitution and other social problems. Where land is taken for mining, forced or mass displacement may result. Traditional collective ownership of land may be disrespected for lack of formal titles, and legal (or corrupt) titles given or sold to mining companies. When mines close, ravaged hills and streams may not be restored, or communities may be left to pay the costs, or restoration may be impossible for decades because of acid drainage.

The economic impact on communities can also be significant. As Albino Barrera points out in his chapter, mining operations in the Global South rarely help the industry generate sustainable wealth in the host country, but instead allow for the formation of extractive-export enclaves. This pattern helps to keep the Global North enriched and the Global South impoverished, while the latter remains on the front lines in coping with climate change.

In many instances, these problems lead to resistance and protests by community members and their defenders. In countries where the rule of law is weak (as in much of the Global South), they may be arrested, jailed, prosecuted on false or trumped-up charges, threatened, harassed, tortured, or killed (see DeJusticia and Business and Human Rights Resource Centre 2015). And in nations beset by war, mining may become a major source of funds to fuel armed conflict, such as "blood diamonds" in Sierra Leone (King n.d.) or coltan and "3TG" (tin, tantalum, tungsten and gold) in the Congo (Heath 2014).

Examples of human rights cases against mining companies

Colombia. The range of human rights issues that can arise at a single mine is illustrated by a recent complaint filed against three of the world's mining giants—BHP, Anglo American, and Glencore. Supported by several other human rights groups, the Global Legal Action Network (GLAN) filed a complaint against the three companies with the National Contact Points (NCPs) in Australia, Switzerland, and the UK (their home countries) in connection with their joint venture ownership of the El Cerrejón mine in Colombia. The complaint alleges "Non-Compliance with the OECD Guidelines for Multinational Enterprises" (GLAN 2021). El Cerrejón is one of the largest open-pit coal mines in the world. GLAN alleges that the mine has caused air and noise pollution and related health impacts, contamination of a river and nearby waterways, interference with the local hydrological system, food insecurity, and an adverse impact on climate change. GLAN further alleges that the three companies have inadequately responded to threats against activists and have displaced indigenous communities, damaged their cultural heritage, and undertaken other activities without their consent.

The complaint adds that the UN Special Rapporteur on Human Rights and the Environment declared El Cerrejón "one of the most disturbing situations" he had learned about (GLAN 2021, 2n3). At this writing, the NCPs have not yet reached conclusions concerning these allegations. Meanwhile, the jointly-owned company announced an agreement in February 2021 with the Wayúu community, by which it would resolve a Colombian court case, reportedly by building a community health center; carrying out an environmental rehabilitation program to include the planting of 250,000 trees; cleaning community facilities and nearby areas over the next five years; and providing technical support for measuring air quality (Reuters 2021). Nonetheless, Glencore has recently brought an international arbitration claim against Colombia over El Cerrejon, and the NCP complaints illustrate the issues of human rights, environment, and conflict that often arise at mines (Bnamericas 2021).

Guatemala. In 2013 a Superior Court of Ontario, Canada, denied a motion to dismiss three lawsuits brought by indigenous Mayans in Guatemala against Canada's HudBay Minerals. Community members had protested a proposed open-pit nickel mine. The Mayan claimants alleged that security personnel working for HudBay's former wholly-owned subsidiaries, allegedly under the control and supervision of HudBay, violated human rights. Among other alleged violations were the killing of a community leader, a shooting, and gang rapes in the vicinity of the proposed mine (*Choc v. Hudbay Minerals Inc.* 2013). While the rapes were committed in 2007, before HudBay acquired the project from another Canadian company in 2008, the killing and shooting allegedly occurred in 2009. Civil litigation remains pending. Meanwhile, in 2021, the mine's former security chief (also formerly a lieutenant general in Guatemala's army) pled guilty in a criminal case to murdering the community leader and shooting another man who was left paralyzed (Waddell Phillips Professional Corporation 2021).

Zambia. In 2019, the United Kingdom Supreme Court permitted a civil suit brought by more than 2,500 Zambians to proceed in London against a British company and its Zambian subsidiary for leaks of acidic material. It also alleged that the company knowingly discharged toxic metals above legal limits into river waters near a copper mine in Zambia (*Vedanta Resources PLC v. Lungowe et al.* 2019). According to the summary by the Court of Appeals, the claimants alleged that they relied on the waterways "as 'their primary source of clean water for drinking, bathing, cooking, cleaning and other domestic and recreational purposes.' It is said that the waterways irrigate crops and sustain livestock [and] are a source of fresh fish and that, in consequence, the waterways are 'of critical importance to [the claimants'] livelihoods and their physical, economic and social well-being'" (*Lungowe et al. v. Vedanta Resources PLC and Konkola Copper Mines PLC*2017). In 2021 the case was settled for an undisclosed sum (Leigh Day 2021).

Mining and *Laudato Si'*

What is the stance of Catholic social teaching toward such alleged abuses? This section considers how the children of God should treat our common home—the earth—as set forth in Pope Francis's 2015 encyclical *Laudato Si'*. Aspects of its teaching relate to the impacts of mining on human rights, the environment, and peace, and it calls for a deeper change of the culture and values that underpin problems associated with mining. *Laudato Si'* teaches that abuse of our planetary home is not an accident but has "ethical and spiritual roots" (§9). It reflects anthropocentrism, individual and societal selfishness, and the myths of markets and technology as saviors. We indulge "the notion that there are no indisputable truths to guide our lives, and hence human freedom is limitless" (§6). We demand rights without corresponding responsibilities. Such a culture is prone to promote lucrative mining without the corresponding human rights and environmental safeguards, genuine community consultation, or respect for communally owned indigenous land. If buyers demand minerals regardless of the real end use value, the market rules. If environmental protection is needed, mining companies either downplay the problem or look to technological solutions, even if they merely mitigate the damage or postpone the day of reckoning. *Laudato Si'* diagnoses the cultural maladies more specifically.

Self-centeredness. Modern culture is beset by "rampant individualism," a "self-centered culture of instant gratification," and a short-term mindset (§161, 181). Closely related is an egoistic attitude that we can plunder Sister Earth at will. In contrast, the creation accounts of the Bible leave "no place for a tyrannical anthropocentrism unconcerned for other creatures" (§68). Francis observes, "Once we lose our humility and become enthralled with the possibility of limitless mastery over everything, we inevitably end up harming society and the environment" (§224).

Technology. This misconception is what allows for the "technocratic paradigm" that operates by prioritizing control over external objects (§106). It lures us into accepting the idea that technology and ingenuity can accomplish unlimited growth, but it is "based on the lie that there is an infinite supply of the earth's goods, and this leads to the planet being squeezed dry beyond every limit." The mining industry is not specifically named here, but falls clearly within the purview of Francis's concerns.

Technology works wonders in medicine and science. In the market, however, the technocratic paradigm leads to a "reductionism which affects every aspect of human and social life," and its products are "not neutral, for they create a framework which ends up conditioning lifestyles and shaping social possibilities along the lines dictated by the interests of certain powerful groups" (§107). The tendency is for the economy to accept "every advance in technology with a view to profit, without concern for its potentially negative impact on human beings" (§109).

Markets. Economic interests such as mining enterprises can trump the common good (§54). While it is supposed to efficiently allocate resources, the market often assigns a value to products that might not reflect their real worth and that can lead to overproduction, which negatively impacts the environment and regional economies. Francis rejects "a magical conception of the market" (§190). The point is not to stifle progress, but to define it in ways other than profit and economic growth. Economic development which does not "respect and promote human rights—personal and social, economic and political, including the rights of nations and of peoples—would not be really worthy of man" (§93). Therefore, decreased growth might be justified in some parts of the world if it means better protection of the environment, traditional cultures, human rights, and peace (§193). This is not saying that mining companies or other growth-focused multinational businesses are inherently evil. "Business is a noble vocation, directed to producing wealth and improving our world" (§129). Mining creates jobs and gives the world materials. Despite those real benefits, accepting decreased growth in the mining sector in some cases could be an antidote to the technocratic paradigm and a market that makes growth and profit goals unto themselves at the expense of other, more important, values.

Environment. The well-being of the environment is one of those values (§23–26). At least three specific environmental problems addressed in *Laudato Si'* relate directly to mining: waste and our "throwaway culture," climate change, and water poverty. The victims of these problems are disproportionately in the Global South. That imbalance is connected to the "disproportionate use of natural resources by certain countries [in the Global North] over long periods of time" (§51). A "throwaway culture" leads to wasteful over-consumption, which worsens climate change and yields more demand for mined resources from the Global South, threatening water sources with pollution and creating greater climate vulnerability in areas where the mining occurs (§20–22).

A true ecological approach, Pope Francis teaches, "*always* becomes a social approach; it must integrate questions of justice in debates on the environment, so as to hear *both the cry of the earth and the cry of the poor*" (§49). This social approach is evident in the church's insistence that, while a right to private property is important, there is always a "social mortgage on all private property" (§93). And the natural environment affected by mining is not private property, but a "collective good, the patrimony of all humanity and the responsibility of everyone" (§95). The same cultural and ethical problems that underlie climate change underlie poverty. Similarly, mining not only harms the environment but sometimes exhausts resources on which communities depend, undermines social structures which shape communal identity, and destroys sacred spaces (§146). We are not faced, then, with "two separate crises, one environmental and one social, but rather with one complex crisis which is both social and environmental," and which demands "an integrated approach to combating poverty, restoring dignity to the excluded, and at the same time protecting nature" (§139).

Certain questions should be asked of any proposed venture (including a mine) to discern whether it will contribute to "genuine integral development. What will it accomplish? Why? Where? When? How? For whom? What are the risks? What are the costs? Who will pay these costs and how?" (§185). Francis also specifically insists that environmental impact assessments should come, not after business plans are already made, but as part of the decision-making process from the beginning. Consensus should be sought among the different stakeholders, "who can offer a variety of approaches, solutions and alternatives," and the local population should have a "special place at the table" (§ 183). Assessments should be transparent enough and honest enough to allow those populations to make truly informed decisions (§184).

The need to act is urgent. "The pace of consumption, waste and environmental change has so stretched the planet's capacity that our contemporary lifestyle, unsustainable as it is, can only precipitate catastrophes" (§161). Sustainable development is a basic question of justice and "intergenerational solidarity" (§159). Uncertainty cannot be permitted to paralyze action. *Laudato Si'* invokes the precautionary principle, enunciated by the 1992 Rio Declaration of the UN Conference on Environment and Development: "where there are threats of serious or irreversible damage, lack of full scientific certainty shall not be used as a pretext for postponing cost-effective measures" to protect the environment (UNGA 1992).

Peace. A spiritual resource for ushering in the kind of cultural reform Francis describes is the cultivation of peace. St. Francis understood, as his namesake observes, "just how inseparable the bond is between concern for nature, justice for the poor, commitment to society, and interior peace" (Francis 2015, §10). *Laudato Si'* explains that inner peace is essential to the conversion that is needed for real cultural change along the lines of integral ecology (§225). Inner peace recognizes that "less is more" (§222). However, mining places exterior or communal peace at risk. As illustrated by the above case of alleged abuses in Guatemala, mining may not only disturb individual and social tranquility, but also often ignites violence or even fuels armed conflict. And indeed, Francis clearly includes concern for peace in the matrix of integral goods that are threatened by ecological indifference and preoccupation with profit over justice and human dignity (§82, 92).

A Way Forward. The cultural values and norms highlighted above suggest that "bold cultural revolution" is needed (§114). While appreciating material progress, we must recover values "swept away by our unrestrained delusions of grandeur." We must not seek to shape the earth solely to our own ends, but instead engage in "responsible stewardship" (§116). We cannot renew our relationship with nature "without a renewal of humanity itself. There can be no ecology without an adequate anthropology" (§118). The challenge is to "to bring the whole human family together to seek a sustainable and integral development," one sensitive to the suffering imposed by environmental degradation, economic inequities, and violence on the poor (§13).

Nor can cultural change do it alone. For changes in culture to succeed, they must be accompanied by changes in laws (§53). But the reverse is also true. If the laws are to bring about "significant, long-lasting effects, the majority of the members of society must be adequately motivated to accept them, and personally transformed to respond" (§211). Moreover, even well-framed laws governing the mining industry are often ineffective because much of the world's mining takes place in nations with weak institutions, little respect for the rule of law, and citizens who lack political power (§142).

But national laws and political power are not enough. Individual countries often face a power imbalance in confronting huge, multinational mining companies. For example, the Anglo-Swiss mining giant Glencore had revenues in 2019 of $215 billion (Mining Technology 2020)—more than the entire GDPs of mining nations such as Bolivia ($41 billion), Guatemala ($77 billion), or Zambia ($23 billion) (World Bank n.d.). It follows that "[e]nforceable international agreements are urgently needed… to impose obligations and prevent unacceptable actions, for example, when powerful companies or countries … offshore polluting industries in other countries" (Francis 2015, §173).

International human rights norms

International human rights norms do not presuppose a belief in the divine. They proclaim freedom, not only of religion, but also of conscience—which includes the right not to believe in any divine power (UNGA 1966a, art. 18.1). They do not presume to denounce the materialistic values of contemporary culture, or to challenge the prevalent paradigm of unending technological progress and economic growth. They are the product, not of a unified epistemology, but of negotiated—or feigned—agreement on fundamental rights among disparate cultures and political systems.

While they do not join Catholic social teaching, or indeed any religion, in reaching for the heavenly, international human rights norms are consistent, as far as they go, with Catholic social teaching. They are centered on human dignity (UNGA 1948, art. 1). They ask states to respect not only civil and political rights (UNGA 1966a), but also economic, social, and cultural rights (UNGA 1966b). They consecrate workers' rights to decent wages and hours, working conditions, and to organize and bargain collectively (UNGA 1948, art. 23; International Labour Organization 2010). They recognize that the right to private property can be subordinated to public purpose (Organization of American States 2011, art. 21.1), albeit with due monetary compensation (art. 21.2). They include the human right to a decent environment (UNGA 2018, §1; African Union 1981, art. 24; Organization of American States 1999, art. 11.1; European Union 2012, art. 37; Association of Southeast Asian Nations 2013, art. 28.f; Community of Latin American and Caribbean States 2018). Some of their instruments embrace the precautionary principle (Organisation for Economic Co-operation and Development 2011, art. VI.4). They call for meaningful consultation with affected communities (UN Office of the High Commissioner for Human Rights 2011, principle 18.a) and, for indigenous peoples, "free, prior and informed consent" before initiating major development projects (UNGA 2007, art. 10, 11.2, 19, 29.2, 32.2). International norms include a right to development, albeit vaguely defined and largely unenforceable (UNGA 1986).

For mining—and indeed for business generally—the central instrument is the UN *Guiding Principles on Business and Human Rights* (UNGPs), adopted by consensus of the UN Human Rights Council in 2011, and now widely incorporated or endorsed by other international and regional instruments, as well as by industry and company policies, and the church (UN Office of the High Commissioner for Human Rights 2011).

The UNGPs rest on three pillars. Pillar I is addressed to states. It reaffirms that they have a "duty to protect" human rights from infringement by third parties, including business (principle 1). This largely restates and particularizes legal duties of states under existing international human rights law.

Pillar II addresses business. It recognizes that business has a "responsibility to respect" human rights (principle 11), which is less demanding than the state's "duty to protect." The responsibility is both negative and positive. Negatively, companies should not violate human rights. Positively, they should adopt formal human rights policies (principle 15, 16); exercise due diligence to anticipate and avoid adverse human rights impacts which they cause or contribute to through their own activities (principle 13, 17); use their leverage to seek to prevent or mitigate adverse human rights impacts arising from their business relationships (principle 13); engage in meaningful consultation with potentially affected groups and other stakeholders (principle 18); monitor and report on the human rights impacts of their activities on a regular basis (principle 20, 21); and address such impacts when they occur (principle 15, 19). Even if the responsibility of business to respect human rights is not (yet) a legal duty, neither is it purely voluntary. It reflects societal expectations of business, which are now articulated, not only in the UNGPs, but in the many governmental and business documents which adopt it (UNGA 2008, §54).

Pillar III is addressed to both states and business. It recognizes the right of victims to remedies for business-related human rights abuses. States have a duty to provide both judicial and non-judicial remedies (UN Office of the High Commissioner for Human Rights 2011, principle 25–28). Where business identifies that it has caused or contributed to an adverse impact, it has a responsibility to provide or cooperate in remediation through legitimate processes (principle 22). Business also has a responsibility to provide non-judicial remedies through genuine, rights-compatible grievance procedures (principle 29–31).

The UNGPs do not call on companies to love their neighbors or renounce materialism. They nonetheless add a welcome degree of practical effectiveness and specificity to Catholic social teaching on human rights and ecology as applied to mining. The principles themselves are relatively precise. For example, principle 16 provides not only that a company should have a "policy commitment" to respect human rights, but more specifically that the policy should: be approved at the most senior level; be informed by expertise; state the company's human rights expectations of personnel, business partners and other parties directly linked to its activities; be publicly available and communicated internally and externally; and be reflected in operational policies and procedures necessary to embed it throughout the company.

The UNGPs lend themselves to implementation by business generally and the mining industry in particular. In 2018 the International Council of Metals and Mining (ICMM), representing 27 of the world's largest mining companies, announced that it had become the first mining industry body to commit its members to implement the UNGPs. The principles are thus to be put in practice by ICMM member companies, which together cover over a quarter of all mined commodities by value.

The UNGPs also signal the business responsibility to respect an array of other international human rights instruments, most of which were originally directed largely or only to states. Principle 12 calls on business to respect the International Covenant on Civil and Political Rights; the International Covenant on Economic, Social and Cultural Rights; the Universal Declaration of Human Rights; and the International Labour Organization's Declaration on Fundamental Principles and Rights at Work. Together these instruments cover a broad spectrum of human rights. Even so, for particular contexts such as mining, principle 12 goes further, recognizing that "[a]dditional standards may be applicable in particular situations, such as the UN Declaration on the Rights of Indigenous Peoples." Among other provisions, the UN Declaration requires that free, prior, and informed consent of indigenous peoples be sought before mining on their lands or using their waters or other resources (UNGA 2007, e.g., art. 32.2).

Formally, the Declaration is not legally binding. Nonetheless, it can be used to assess whether business meets its responsibility to respect human rights. It also informs interpretation of treaties which are legally binding on states. For example, International Labour Organization Convention 169 requires "free and informed consent" before indigenous peoples are relocated (International Labour Organization 1989, art. 16.2). Although joined by only 23 states, they include such important mining countries as Bolivia, Brazil, Chile, Mexico, and Peru.[2]

Regional human rights treaties have also been interpreted to require free, prior, and informed consent. In the Americas, nearly all Latin American countries are parties to the American Convention on Human Rights. The Convention has been interpreted by the Inter-American Court of Human Rights (2012, §177–211) to mandate free, prior and informed *consultation* with indigenous peoples. Moreover, in cases involving major impact projects within indigenous territory, "the State has a duty, not only to consult with [them]…, but also to obtain their free, prior, and informed *consent*, according to their customs and traditions" (Inter-American Court of Human Rights 2007, §134). In Africa, the African Court of Human and Peoples' Rights (2017, §131) has held that States have at least a duty of "prior consultation" (and perhaps consent) before expelling an indigenous people from its land. These injunctions to consult versus obtain consent reflect varying legal interpretations of the force such regulations hold. The above sources refer those regulations to states, but the UNGPs make companies, including mining companies, responsible to use the leverage they may have to try to ensure that genuine consultation takes place and that, at minimum, free, prior, and informed consent is sought.

The foregoing discussion of norms from the UNGPs that are applicable to mining is not exhaustive. Still, it provides starting points for practitioners of Catholic peacebuilding who wish to bring widely supported secular norms to bear in the defense of human dignity, planetary survival, the common good, and peace.

International norms on peace

With relatively few but nonetheless important exceptions,[3] international law does not generally address the kind of interior peace of which St. Francis wrote, which is likely to be damaged by the kinds of human rights and environmental abuses associated with mining. International law mainly addresses the right to external peace—the absence of armed conflict. The UN General Assembly's Declaration on the Right to Peace provides, "Everyone has the right to enjoy peace such that all human rights are promoted and protected and development is fully realized" (UNGA 2016, art. 1). The human rights provisions of the Declaration address not only international wars and internal armed conflicts, but arguably also civil disturbances, such as those arising from mining disputes (art. 2). By themselves, this and earlier UN declarations on peace are not legally binding. However, they can be politically and diplomatically influential. They may also be used judicially to interpret treaties (such as the UN Charter).

The UNGPs are also not legally binding but provide guidance to business, including mining companies, operating in conflict zones:

> '[C]onflict-affected areas' may increase the risks of enterprises being complicit in gross human rights abuses committed by other actors (security forces, for example). Business enterprises should treat this risk as a legal compliance issue, given the expanding web of potential corporate legal liability arising from extraterritorial civil claims, and from the incorporation of the provisions of the Rome Statute of the International Criminal Court in jurisdictions that provide for corporate criminal responsibility. In addition, corporate directors, officers and employees may be subject to individual liability for acts that amount to gross human rights abuses. In complex contexts such as these, business enterprises should ensure that they do not exacerbate the situation. (UN Office of the High Commissioner for Human Rights 2011, principle 23 commentary)

Aside from these human rights instruments, most international norms on peace in the UN Charter, international humanitarian law, and international criminal law do not usually apply to mining disputes, for two main reasons. First, most international law on resort to war is restricted to wars between nations, which are rarely triggered by mining disputes.[4] Second, international laws mandating humane conduct of internal armed conflicts between governments and insurgents would apply to conflicts tied to mining only if they involve a high threshold of military organization, discipline, and violence, as with rebel groups in the Democratic Republic of the Congo, though not the riots or sporadic violence typically involved in mining disputes (International Criminal Court 2011, art. 8.2.d).

However, international law does not ignore communities victimized by violence in mining disputes. International human rights law applies. It protects, for example, the rights to life and bodily integrity, as well as freedoms of speech, association, and peaceful assembly (UNGA 1966a, art. 6, 7, 10, 18, 19, 22). In cases before international human rights bodies, advocates may use the UN Declaration on the Right to Peace "within and between societies," and the UNGPs on risks in conflict zones, to help interpret human rights norms in the context of violent mining disputes.

Combining for peace and the common good

While Catholic social teaching is more ambitious and aspirational, international human rights norms are more concrete and immediately practical (even if narrower). In addressing mining, Catholics need not choose between the two. The church will not and should not retreat from its commitments inspired by the Gospel. At the same time, as *Laudato Si'* teaches, the task before us is so immense that Catholics need to reach out to all people of good will to curb the calamities to which mining often contributes.

That means reaching out, beyond even other religious communities, to secular activists, victims of abuses, NGOs, human rights and environmental lawyers, courts, governments, the media—and mining companies, with whom the church has dialogued in recent years (Francis 2019). Outreach is best communicated through language readily understood by others, and by invoking norms they embrace, so long as those norms are consistent with Catholic social teaching. In addressing mining, the Catholic community would be well-advised not to limit itself to Catholic doctrine. The church should endeavor to ensure that its recommendations on mining to governments, business, and society are consistent with, and capture the essentials of, relevant international human rights norms, including but not limited to the widely endorsed UNGPs. Among other things, this means becoming acquainted with the content of the UNGPs. In general, the state's "duty to protect" and the business's "responsibility to respect" human rights (including environmental rights and the right to peace) are eminently consistent with Catholic social teaching. If in any instance the church believes that it must depart from an aspect of the UNGPs, the need for such a departure should be carefully reasoned and publicly articulated. Outreach has practical value. Even if it may not meet all the broad spiritual objectives of the church, it adds social and political heft by uniting the persuasive powers of both religious and secular norms and actors against the power of multinational mining companies. It thereby concretely advances Catholic social principles on human rights, the environment, and peace. The church can and should unite its voice with victims and with those who advocate putting secular norms into practice, as opposed to merely articulating them in empty promises.

In reaching out, the church can bring to bear its broader perspective. For example, cost-benefit analyses of mining projects should not be limited to impacts in the local community, but should be expanded to take into account the social utility, or lack thereof, of the end for which the mining is undertaken. It is one thing to expose local residents to some degree of risk in order to mine precious metals essential for computer networks in hospitals. It is quite another to ask them to sacrifice in order to mine gold or silver for luxury trinkets. If communities near a mine are to be asked to face some degree of risk, it should at the very least be for morally justifiable ends. This point, not addressed by secular international human rights norms, is a spiritual truth taught by *Laudato Si'* and its critique of mining in light of larger cultural maladies of consumerism, materialism, and waste.

However, assessing end use may be difficult in practice. If a mineral is mined for sale on the world market, there may be no way in advance to know or to value its eventual end use. It may eventually be sold as jewelry or used for heart monitors. Or a mining company may know the end uses but refuse to disclose them. Or there is the conundrum of how to assess mining which damages the environment when the minerals are to be used for renewable energy technology to mitigate climate change, as discussed by Holden and Montevecchio later in this book. To overcome such practical obstacles, the church could learn both positive and negative lessons from secular norms and procedures on mining due diligence (European Parliament and the Council of the European Union 2017), transparency (Extractive Industries Transparency Initiative n.d.), disclosure (US Congress 2010, sec. 1502.b), reporting (UK Parliament 2015, sec. 52), and certifications (Kimberley Process n.d.). The church does not need to reinvent wheels. Secular experience may either point the way to solutions, signal where they are elusive, or clarify when certain approaches are unworkable. On the other hand, without advocacy by the church and others of broader vision, international and national laws might continue to permit mines to be licensed without regard to end uses. Just as Catholic social teaching can learn from international human rights law, so too international law can learn from Catholic principles.

Laudato Si' further teaches that necessary cultural change cannot fully succeed without legal change. This means that the church should learn about existing national mining laws and advocate reforms. In Guatemala in the early 2000s, for example, the church played a key role in developing a draft reform of the mining law (Holden and Jacobson 2009). Although the draft was not adopted by the country's lawmakers, the church stood—and was seen to stand—with the people and on the right side of history.

Elsewhere church campaigns on mining may meet with greater success, as seen in two cases from this book. As outlined by Andrés McKinley, in El Salvador church support was important to a ban on mining, which the government later successfully defended in an international arbitration brought by mining companies (Palumbo and Malkin 2017; Malkin 2016). The ban on mining "marked the culmination of a multi-year campaign in which the archbishop of San Salvador and the Jesuit-run Central American University played a major role" (Palumbo 2017). And in the chapter on Peru, José Bayardo Chata Pacoricona describes several court successes arguing against mining development and active pursuit of varied legal actions on behalf of indigenous and other marginalized communities.

As the contrasting outcomes in the neighboring countries of El Salvador and Guatemala suggest, national laws alone may not always prevail over the wealth and power of large mining companies. Since the operations of multinational mining enterprises extend beyond the reach of any one nation it is important, as recognized by *Laudato Si'*, to develop global regulatory norms and institutions with the power to impose sanctions. For example, without endorsing any particular text, the Holy See has repeatedly endorsed the concept of a legally binding treaty on business and human rights to regulate transnational business, including mining companies. Citing *Laudato Si'*, the Holy See in 2017 told the UN intergovernmental group working on a proposed treaty that:

> [I]t is important to recognize that there are good reasons why international law might devote specific attention to transnational corporations and in particular their accountability for human rights abuses. An international legal instrument has the potential to make corporations criminally, civilly, and administratively liable, while guaranteeing the protection of human rights, providing access to judicial remedy, and adding an important tool for accountability. (Jurkovič 2017, n.p.)

These are examples of how the church can work on mining issues with other faith and secular communities: through outreach, in campaigns to improve national laws, and in shaping and invoking international laws and institutions. The church must of course continue its indispensable spiritual and social teaching. However, if the church is to overcome injustice in mining, a bridge to persons of goodwill can be built by learning the language and using the norms of international human rights. Neither Catholic social teaching nor international norms suffice by themselves to shield human rights, the environment, and peace, from the adverse impacts of mining. What the church needs is an approach grounded in the Gospel but also informed by international human rights law.

Notes

1 Compare, for example, Green America (n.d.) with Totty (2008).
2 Information on ratifications of the *Indigenous and Tribal Peoples' Convention* can be found at https://www.ilo.org/dyn/normlex/en/f?p=NORMLEXPUB:11300:::NO:11300:P11300_INSTRUMENT_ID:312314:NO.
3 For example, mental pain and suffering may be included as an element of monetary reparations for gross violations of human rights (see Inter-American Court of Human Rights 2003, §260–66; International Criminal Court 2011, art 7.1.k).

4 One exception is armed conflict between insurgent groups supported by Rwanda in the Democratic Republic of the Congo (see Schneider 2012).

References

African Court on Human and People's Rights. 2017. *African Commission on Human and Peoples' Rights v. Kenya*. Judgment of May 26, 2017. https://www.escr-net.org/caselaw/2017/african-commission-human-and-peoples-rights-v-republic-kenya-acthpr-application-no.

African Union. 1981. *African Charter on Human and Peoples' Rights*. OAU Doc. CAB/LEG/67/3. June, 1981. https://www.achpr.org/legalinstruments/detail?id=49.

Association of Southeast Asian Nations. 2013. *ASEAN Human Rights Declaration*. https://www.asean.org/storage/images/ASEAN_RTK_2014/6_AHRD_Booklet.pdf.

Bnamericas. 2021. "Glencore files new arbitration claims against Colombia." June 3, 2021. https://www.bnamericas.com/en/news/glencore-files-new-arbitration-claims-against-colombia.

Brown, Catherine. 2016. "State-Owned Copper Mining: Climate Change vs. Country Development." Harvard Business School. November 4, 2016. https://digital.hbs.edu/platform-rctom/ submission/state-owned-copper-mining-climate-change-vs-country-development/.

Choc v. Hudbay Minerals Inc. 2013. ONSC 1414. Ontario. July 22, 2013.

Community of Latin American and Caribbean States. 2018. *Regional Agreement on Access to Information, Public Participation and Justice in Environmental Matters in Latin America and the Caribbean* (Escazú Agreement). March 4, 2018. https://repositorio.cepal.org/bitstream/handle/11362/43583/1/S1800428_en.pdf.

DeJusticia and Business and Human Rights Resource Centre. 2015. *Digging Deeper: The Human Rights Impacts of Coal in the Global South*. Edited by Krizna Gomez and Gregory Regaignon. Creative Commons: November, 2015. http://media.wix.com/ugd/c04a21_3e6e58c1b5804082bf31ac72699d0c53.pdf.

European Parliament and the Council of the European Union. 2017. "Laying Down Supply Chain Due Diligence Obligations for Union Importers of Tin, Tantalum and Tungsten, Their Ores, and Gold Originating from Conflict-Affected and High-Risk Areas." Regulation (EU) 2017/821. May 17, 2017. https://eur-lex.europa.eu/legal-content/EN/TXT/?uri=CELEX%3A32017R0821.

European Union. 2012. *Charter of Fundamental Rights of the European Union*. 2012/C 326/02. October 26, 2012. https://eurlex.europa.eu/eli/treaty/char_2012/oj.

Extractive Industries Transparency Initiative. n.d. "Who We Are." Accessed March 9, 2021. https://eiti.org/who-we-are.

Francis. 2015. *Laudato Si'*. http://www.vatican.va/content/francesco/en/encyclicals/documents/papa-francesco_20150524_enciclica-laudato-si.html.

Francis. 2019. "Address to Participants at the Meeting Promoted by the Dicastery for Promoting Integral Human Development on the Mining Industry." May 3, 2019. http://www.vatican.va/content/francesco/en/speeches/2019/may/documents/papa-francesco_20190503_incontro-industria-mineraria.html.

Glendon, Mary Ann. 1998. "Knowing the Universal Declaration of Human Rights." *Notre Dame Law Review* 73, no. 5: 1153–1190.

GLAN (Global Legal Action Network). 2021. *Non-Compliance with the OECD Guidelines for Multinational Enterprises*. January 21, 2021. https://media.business-humanrights.org/media/documents/Full_text_complaint_before_OECD_by_GLAN.pdf.

Green America. n.d. "10 Reasons to Oppose Nuclear Energy." Accessed March 8, 2021. https://www.greenamerica.org/fight-dirty-energy/amazon-build-cleaner-cloud/10-reasons-oppose-nuclear-energy.

Greenpeace Australia Pacific. 2017. "Steeling the Future: The Truth Behind Australian Metallurgical Coal Exports." June 2017. https://www.greenpeace.org.au/wp/wp-content/uploads/2017/06/280517-GPAP-Steeling-the-Future-Report-LR.pdf.

Griffin, Oliver, and Luis Jaime Acosta. 2020. "Colombia's Mining Future Tied to Metals Not Coal, Minister Says." *Reuters*, October 2, 2020. https://www.reuters.com/article/us-colombia-energy/colombias-mining-future-tied-to-metals-not-coal-minister-says-idUSKBN26N2HK.

Heath, Nick. 2014. "How Conflict Minerals Funded a War That Killed Millions, and Why Tech Giants Are Finally Cleaning Up Their Act." *Tech Republic*, March 31, 2014. https://www.techrepublic.com/article/how-conflict-minerals-funded-a-war-that-killed-millions/.

Holden, William N., and R. Daniel Jacobson. 2009. "Ecclesial Opposition to Nonferrous Mining in Guatemala: Neoliberalism Meets the Church of the Poor in a Shattered Society." *The Canadian Geographer* 53, no. 2: 145–164.

ICMM (International Council of Mining and Metals). 2018. "ICMM First Industry Body to Commit to UN Guiding Principles on Business and Human Rights." November 26, 2018. https://www.icmm.com/en-gb/news/2018/un-human-rights-pr.

Inter-American Court of Human Rights. 2003. *Case of Myrna Mack Chang v. Guatemala*. Judgment of November 25, 2003. https://www.corteidh.or.cr/docs/casos/articulos/seriec_101_ing.pdf.

Inter-American Court of Human Rights. 2007. *Case of the Saramaka People v. Suriname*. Judgment of November 28, 2007. https://www.corteidh.or.cr/docs/casos/articulos/seriec_172_ing.pdf.

Inter-American Court of Human Rights. 2012. *Case of the Kichwa Indigenous People of Sarayaku v. Ecuador*. Judgment of June 27, 2012. https://corteidh.or.cr/docs/casos/articulos/seriec_245_ing.pdf.

International Criminal Court. 2011. *Rome Statute of the International Criminal Court*. https://www.icc-cpi.int/resource-library/documents/rs-eng.pdf.

International Energy Agency. 2020. *Coal 2020: Analysis and Forecast to 2025*, Fuel Report. December 2020. https://www.iea.org/reports/coal-2020.

International Labour Organization. 1989. *Indigenous and Tribal Peoples' Convention*. C169. https://www.ilo.org/dyn/normlex/en/f?p=NORMLEXPUB:12100:0::NO::P12100_ILO_CODE:C169.

International Labour Organization. 2010. *Declaration on Fundamental Principles and Rights at Work*. June 18, 1998 (revised June 15, 2010). https://www.ilo.org/declaration/lang--en/index.htm.

Jurkovič, Ivan. 2017. "Statement by Archbishop Ivan Jurkovič, Permanent Observer of the Holy See to the UN and Other International Organizations in Geneva." October 25, 2017. https://www.ohchr.org/Documents/HRBodies/HRCouncil/WGTransCorp/Session3/OralInterventions/HolySee-Subject5.LegalLiability.pdf.

Kimberly Process. n.d. "What is the Kimberly Process?" Accessed March 9, 2021. https://www.kimberlcyprocess.com/en/what-kp.

King, Hobart M. n.d. "Blood Diamonds—Conflict Diamonds—What is the Kimberley Process?" Accessed March 8, 2021. https://geology.com/articles/blood-diamonds.shtml.

Leigh Day. 2021. "Legal Claim by More Than 2,500 Zambian Villagers in a Case Against Vedanta Resources Limited." January 19, 2021. https://www.leighday.co.uk/latest-updates/news/2021-news/legal-claim-by-more-than-2-500-zambian-villagers-in-a-case-against-vedanta-resources-limited/.

Lungowe et al. *v. Vedanta Resources PLC and Konkola Copper Mines PLC*. 2017. EWCA Civ 1528. November 2017.

Malkin, Elizabeth. 2016. "El Salvador Wins Dispute Over Denying a Mining Permit." *New York Times*, October 14, 2016. https://www.nytimes.com/2016/10/15/world/americas/salvador-mining-dispute.html.

Mining Technology. 2020. "Top Ten Metals and Mining Companies in 2020." Updated October 12, 2020. https://www.mining-technology.com/features/top-ten-metals-and-mining-companies-in-2020/.

Organisation for Economic Co-operation and Development. 2011. *OECD Guidelines for Multinational Enterprises*. http://www.oecd.org/daf/inv/mne/48004323.pdf.

Organization of American States. 1999. *Additional Protocol to the American Convention on Human Rights in the Area of Economic, Social and Cultural Rights*. A-52. November 16, 1999. https://www.refworld.org/docid/3ae6b3b90.html.

Organization of American States. 2011. *American Convention on Human Rights*. https://www.oas.org/dil/treaties_b-32_american_convention_on_human_rights.pdf.

Palumbo, Gene. 2017. "Spurred by Catholic Leaders, El Salvador Becomes First Nation to Ban Mining." *America*, April 3, 2017. https://www.americamagazine.org/politics-society/2017/04/03/spurred-catholic-leaders-el-salvador-becomes-first-nation-ban-mining.

Palumbo, Gene, and Elizabeth Malkin. 2017. "El Salvador, Prizing Water Over Gold, Bans All Metal Mining." *New York Times*. March 29, 2017. https://www.nytimes.com/2017/03/29/world/americas/el-salvador-prizing-water-over-gold-bans-all-metal-mining.html.

Pooler, Michael. 2019. "Cleaning Up Steel Is Key to Tackling Climate Change." *Financial Times*, January 1, 2019. https://www.ft.com/content/3bcbcb60-037f-11e9-99df-6183d3002ee1.

Reuters. 2021. "Coal Miner Cerrejon Says Reached Deal With Indigenous Community on Environment, Health." February 26, 2021. https://www.reuters.com/article/us-cerrejon-colombia/coal-miner-cerrejon-says-reached-deal-with-indigenous-community-on-environment-health-idUSKBN2AQ2W7.

Schneider, Mark. 2012. "Examining the Role of Rwanda in the DRC Insurgency." Testimony before the U.S. House Committee on Foreign Affairs Subcommittee, September 19, 2012. https://www.crisisgroup.org/africa/central-africa/democratic-republic-congo/examining-role-rwanda-drc-insurgency.

Sepanek, Eric. 2012. "Top 6 Common Uses for Gold." Scottsdale Bullion & Coin. May 30, 2012. https://www.sbcgold.com/blog/top-6-common-uses-for-gold/.

Tomasi, Silvano. 2014. "Statement by Archbishop Silvano M. Tomasi, Permanent Observer of the Holy See to the United Nations and Other Specialized Agencies in Geneva." June 11, 2014. http://www.vatican.va/roman_curia/secretariat_state/2014/documents/rc-seg-st-20140611_tomasi-diritti-umani_en.html.

Totty, Michael. 2008. "The Case For and Against Nuclear Power." *Wall Street Journal*, June 30, 2008. https://www.wsj.com/articles/SB121432182593500119.

UK Parliament. 2015. *UK Modern Slavery Act 2015*. C. 30. March 26, 2015. https://www.legislation.gov.uk/ukpga/2015/30/contents/enacted.

UNGA (UN General Assembly). 1948. *Universal Declaration of Human Rights*. A/RES/217(III). December 10, 1948. https://www.un.org/en/development/desa/population/migration/generalassembly/docs/globalcompact/A_RES_217(III).pdf.

UNGA (UN General Assembly). 1966a. *International Covenant on Civil and Political Rights*. A/RES/2200A(XXI). December 16, 1966. https://www.un.org/en/development/desa/population/migration/generalassembly/docs/globalcompact/A_RES_2200A(XXI)_civil.pdf.

UNGA (UN General Assembly). 1966b. *International Covenant on Economic, Social and Cultural Rights*, A/RES/2200A(XXI). December 16, 1966. https://www.un.org/en/development/desa/population/migration/generalassembly/docs/globalcompact/A_RES_2200A(XXI)_economic.pdf.

UNGA (UN General Assembly). 1986. *UN Declaration on the Right to Development*. A/RES/41/128. December 4, 1986. https://undocs.org/A/RES/41/128.

UNGA (UN General Assembly). 1992. "Rio Declaration on Environment and Development." A/CONF.151/26 (Vol. I). August 12, 1992. https://www.un.org/en/development/desa/population/migration/generalassembly/docs/globalcompact/A_CONF.151_26_Vol.I_Declaration.pdf.

UNGA (UN General Assembly). 2007. *UN Declaration on the Rights of Indigenous Peoples*. A/RES/61/295. September 13, 2007. https://undocs.org/A/RES/61/295.

UNGA (UN General Assembly). 2008. *Report of the Special Representative of the Secretary-General on the Issue of Human Rights and Transnational Corporations and Other Business Enterprises, John Ruggie*. Human Rights Council Session 8. A/HRC/8/5. April 7, 2008. https://www.undocs.org/A/HRC/8/5.

UNGA (UN General Assembly). 2011a. *Human Rights and Transnational Corporations and Other Business Enterprises*. Human Rights Council Session 17. A/HRC/RES17/4. June 16, 2011. https://documents-dds-ny.un.org/doc/RESOLUTION/GEN/G11/144/71/PDF/G1114471.pdf?OpenElement.

UNGA (UN General Assembly). 2011b. *Report of the Special Representative of the Secretary-General on the Issue of Human Rights and Transnational Corporations and Other Business Enterprises, John Ruggie*. Human Rights Council Session 17. A/HRC/17/31. March 21, 2011. https://undocs.org/en/A/HRC/17/31.

UNGA (UN General Assembly). 2016. *UN Declaration on the Right of Peoples to Peace*. A/RES/71/189. December 19, 2016. https://undocs.org/en/A/RES/71/189.

UNGA (UN General Assembly). 2018. *Human Rights and the Environment*. Human Rights Council Session 37. A/HRC/RES/37/8. March 22, 2018. https://undocs.org/en/A/HRC/RES/37/8.

UN Office of the High Commissioner for Human Rights. 2011. *UN Guiding Principles on Business and Human Rights*. https://www.ohchr.org/Documents/Publications/GuidingPrinciplesBusinessHR_EN.pdf.

US Congress, House. 2010. *Dodd-Frank Wall Street Reform and Consumer Protection Act*. HR 4173, 111th Congress. July 21, 2010. https://www.congress.gov/bill/111th-congress/house-bill/4173/text.

UITA (US International Trade Administration). 2019a. "Bolivia—Mining." In *Bolivia Country Commercial Guide*. Last updated July 12, 2019. https://www.export.gov/apex/article2?id=Bolivia-Mining.

UITA (US International Trade Administration). 2019b. "Mexico—M. Mining and Minerals." In *Mexico Country Commercial Guide*. Last updated October 17, 2019. https://www.export.gov/apex/article2?id=Mexico-Mining-and-Minerals.

UITA (US International Trade Administration). 2020. "Mining Equipment and Machinery." In *Peru Country Commercial Guide*. Last updated October 11, 2020. https://www.trade.gov/country-commercial-guides/peru-mining-equipment-and-machinery.

UITA (US International Trade Administration). "Mining." 2021. In *Chile Country Commercial Guide*. Last updated January 16, 2021. https://www.trade.gov/knowledge-product/chile-mining-and-minerals.

Vedanta Resources PLC v. Lungowe et al. 2019. UKSC 2017/0185. April 10, 2019.

Waddell Phillips Professional Corporation.2021. "Head of Security at Hudbay Minerals' Former Guatemalan Mine Pleads Guilty to Killing Community Leader and Shooting Bystander." *YahooFinance*, January 6, 2021. https://ca.finance.yahoo.com/news/head-security-hudbay-minerals-former-215000236.html.

The World Bank. n.d. "GDP (current US $)." Accessed March 8, 2021. https://data.worldbank.org/indicator/NY.GDP.MKTP.CD.

Wu, Guohua, and Yingnan Li. 2018. "Mining in China: Overview." *Thomson Reuters Practical Law*, May 1, 2018. https://uk.practicallaw.thomsonreuters.com/w-0111348?transitionType=Default&contextData=(sc.Default)&firstPage=true.

11 Development as depth: towards a theology of integral human development

Clemens Sedmak

In the first book of Samuel, we find the colorful description of a thought-provoking scene: the young shepherd, David, is willing to fight Goliath; King Saul gives David his armor and helmet, but David "tried in vain to walk, for he was not used to them" (1 Sam. 17:38–39a). David is too small for the armor, probably the most precious and heaviest of the time.

The image is quite powerful if we translate it into the present situation: we have built huge armors of technological possibilities and agency, but have not grown to match it; we are like moral and spiritual dwarfs in the armor of giants. We still study Aristotle's *Nicomachean Ethics*. His contributions to science have lost much of their relevance and plausibility, while we have not made enough progress on the grounds of morality to render his moral observations obsolete. There is a disconnect between technological and moral progress, between economic and spiritual growth. In light of the challenges we face with the practices of mining, we have to ask ourselves some deep moral questions. We cannot do without the fruits of mining in our present lives, but we might have to rethink some of the parameters guiding the industry. This concerns especially decision-making processes and the interaction between mining activities and local contexts. Technological progress may help, but it is not the main issue. To quote a line from Holden and Montevecchio's chapter for this book: "the idea of removing *less* from the earth and, instead, getting *more* out of what has already been extracted or else changing consumption patterns altogether are never considered." These challenges touch upon moral and spiritual questions—and the deliberation of ends. We have, as Pope Francis's (2015, §105) encyclical *Laudato Si'* states, "not been trained to use power well." "We have too many means and only a few substantial ends" (§203).

A theology of integral human development (IHD) is committed to bridging this gap between means and ends, between external progress and inner growth; it is committed to integrating our technological possibilities into our moral and spiritual identity. This article will develop a theology of IHD with a special emphasis on the mining industry.

DOI: 10.4324/9781003094272-11

Integral human development

A simple understanding of IHD is a short formula: the development of the *whole* person and the development of *each* person. This deceptively simple characterization dates back to 1967, when Paul VI, in *Populorum Progressio*, wrote about "authentic human development." Development "cannot be restricted to economic growth alone. To be authentic, it must be well rounded; it must foster the development of each person and of the whole person" (§14). This concept can be translated into two imperatives: Do not leave anyone behind! Make sure that each dimension of the person counts!

These two imperatives of IHD correspond to the notion of the common good and the notion of human dignity. The common good refers to the flourishing of a community as a community on the basis of the flourishing of each of its members. It is an aspirational concept that can never be fully met. But it expresses the challenge like a thorn in the flesh of social reality: to leave no one behind. Seen in this light, the common good is distinguished from a utilitarian approach that seeks the *greatest* good of the *greatest* number. The common good comes with special attention to the most disadvantaged members of a community and sometimes follows the dynamics of the parable of the lost sheep: to "leave the ninety-nine on the mountains and go in search of the one that went astray" (Mt. 18:12). A commitment to IHD implies a commitment to special consideration for the least privileged and least powerful members of a community; in the case of the mining industry, these may be the local population. Especially in a situation of "big business," the reminder that each person's dignity counts may be called for.

The central notion of IHD is human dignity; it refers to an understanding of a dignified life, a life that corresponds to the dignity of the human person. This idea reflects a particular anthropology. It comes with an image of the human person as a multi-layered being that cannot be reduced to a "one-dimensional" existence, just as Herbert Marcuse (2002 [1964]) had described a one-sided and reductionist approach to development. The human person cannot live without bread, but does not live by bread alone. There is an important aspect of "being beyond having" that requires dignity-sensitive development to be human-centered. Human-centered development, in the words of Denis Goulet (1995, 6–7), refers to a primacy of being over having: "Societies are more human, or more developed, not when men and women '*have* more' but when they are enabled to '*be* more.' The main criterion of development is not increased production or material wellbeing but qualitative human enrichment."

Even though we cannot make unrealistic claims, we can ask the mining industry for a systematic "moral cost" analysis of their economic activities. We can ask for the identification of entry points for moral burdens.

An exercise of mapping the crucial moral challenges is a necessary step in building a morally sensitive, more humane economy. This concern has been at the basis of an understanding of development as integral.

The idea that IHD is the development of each person and the whole person was inspired by Joseph Lebret, a French Dominican and economist, who worked with sea fisheries in France. There, he observed the negative effects of certain economic developments on local workers, notably exploitation by a foreign industry (Bosi 2012, 253). Lebret coined the term "human economy," i.e., an economy that would be "favourable to human development," to "a fully human life," as he wrote in his 1954 essay "Économie et Humanisme" (quoted in Keleher 2019, 29). A "fully human life" is more than a provision of basic goods but is rooted in a sense of *compassion*. Compassion is "an existential fellowship with every man [person] who strives to unite the world under a common destiny and to collectively create the structures that the realization of this aspiration calls for. To love is to identify oneself with one's neighbour, with all men [persons], and to create with them the conditions for their self-fulfilment" (Cosmao 1970, 68).

This lofty language can be translated into corporate social responsibility of mining corporations and their sincere engagement with local communities. Raymond Offenheiser's chapter has pointed to some encouraging developments with regard to indigenous rights or the establishment of consent-oriented processes. There have been clearly intentional efforts to pursue a more inclusive approach that takes non-material aspects such as rights into consideration. The idea of a human-centered development committed to an idea of a full human life and the principles of the common good and human dignity can also be translated into an approach in development ethics (see Keleher 2017; 2019). IHD ethics is an important contribution to a way of thinking about development that takes the dignity and the social nature of the human person seriously. The task we face is not ethics but theology. What is specific about a "theology of IHD"?

A theological perspective

The following reflections offer a "niche product," a general framework for a theology of IHD that can be applied to mining. Since the normative discourse on mining has been spearheaded by ethics rather than theology, I would like to show how theology can add a distinctive voice. Ultimately, the theological perspective is an appeal to reflecting on the ultimate ends. It will not be enough to offer a moral justification for particular means; the fundamental question about the first and last goals have to be raised. The "selling point" of theology is the finiteness of life and resources. In other words, there are natural limits to what can be extracted and all those involved in the industry will die one day. This is the entry point for theological considerations.

Some people believe in an afterlife, others do not. This difference seems to be comparable to matters of taste, but the implications reach far deeper. The responses to sensitive questions like beginning and end of life issues, the state of suffering, the meaning of terms like "success," "justice," "happiness," or "flourishing" are profoundly affected by this distinction; so are key aspects of the mining industry such as "profit," "maximization," and "shareholder values." This difference is fundamental for the genuine place of theology in intellectual and academic discourses. This difference decides whether "the world" will have the "first word" and "last word," or whether theological values from "beyond the world" will. Wittgenstein famously stated in his *Tractatus Logico-Philosophicus* (1922, 6.41) that "in the world everything is as it is and happens as it does happen." Similarly, we could say that mining operations happen as they happen, following built-in mechanisms of technologically supported maximization of profit. Normative questions enter from a point of view that does not take the factual as the ultimate norm. Theology specifically asks questions from "outside of the world."

Theology anchors our reflective processes. If we compare our intellectual efforts to a performance on a stage, theology provokes the question of which play we are in. To whom are we talking? From where are we talking? What is the drama in which we are participating? Let us call these three questions the audience question, the position question, and the script question. The audience question: who is listening, who is watching? Candidates for an answer would include contemporaries, future generations, but also ancestors and God. Who will be judging us? The position question: From where are we speaking? Are we speaking from the position of a creature? Are we inhabiting the space of a mortal or an immortal being? The script question: What is the point of the drama we are in? Who has the power to shape the drama, especially the beginning and the end? What is our role on the stage of life?

The project of developing a "theology of IHD" is different from an "ethics of IHD." There is an explicit recognition of audience, position, and script in theology, the recognition of a particular anchoring of our reflection. This anchoring is, of course, always contextual. Theology is not a homogenous country, but many different provinces and regions. There are many "theologies of mining," acknowledging local realities, and local challenges. Conflicts are always entry points for normative considerations, and conflicts always have a particular profile and a local history.

All theologies have this commitment to a specific way of anchoring our thoughts, a commitment to a sphere "beyond the world" that guides and frames our intellectual engagement. This anchoring points to the question of first and last values. If people are tempted to dismiss theology as irrelevant, I would point to three things: (a) the question of first and last values is an essential one that cannot be ignored on an existential level; (b) there is the *possibility* that the story told by religions and reflected on by theologies is true; and (c) the vast majority of people on this planet have some kind of religious affiliation with an either implicit or explicit theology to accompany their religious beliefs and practices. This is very much in line with William James's (1985 [1902]) approach in the *Varieties of Religious Experience*. What can theology offer to the discourse on IHD?

An integral analysis and the foundational text of *Laudato Si'*

Laudato Si' (Francis 2015) is an important and influential document that lays the ground for a theology of IHD. It has been dealt with in this book in the chapter written by Anne Floerke Scheid and Daniel P. Scheid. In their contribution the authors have shown how *Laudato Si'*, with its message on integral ecology, is relevant for an understanding of an ecological just peace applicable to mining. *Laudato Si'* has contributed to a deepening of the concept of IHD with an ecological perspective (§141) and with a long-term perspective that explicitly considers future generations (§159).

If we take an integral view to mean a holistic approach that brings separate aspects into a shared whole, we can understand the integral and integrating character of *Laudato Si'* in many ways. The whole document develops the thesis that everything is connected (§16, 42, 70, 92, 117, 138, 240). As a consequence, the world's problems cannot be analyzed or explained in isolation (§61; cf. §110, 131). This integral view comes with the obligation to think in terms of interdependence, "one world with a common plan" (§164). The integral view, we could say, is an expression of and leads to an ethics of thinking. Let me offer a reference point for a more integral approach.

An integral analysis of the mining industry would take a closer look at the dynamics of "extracting." Extracting is an action that removes something from something else. As Tobias Winright observed in his chapter, it also has the semantic nuance of "force" (if not "violence"): "X extracted information from Y." Here, "extracting" is close to "pulling out forcibly." Etymologically speaking, "extracting" is based on "ex" ("out," "out of") and "trahere" ("draw"). This gives us a first sketch of the key concepts at stake: (a) Extracting is an action (and not an event) that follows agent-causality and intentionality; (b) extracting requires "force;" (c) extracting changes a situation by creating a new ontological state: B is separated from A, changing the state of A as well as the state of B. A basic analysis of "extract," then, could look like this:

X extracts B from A in C through M because of R.

X is the agent (the subject), B is the extracted material, A is the source of the extracted material, C is the context of the operation, M is the mode of extraction (method, means), R is the reason or set of reasons ("why").

The different elements of this analysis are interconnected. If we apply this simple analysis to mining we might see that the agent is normally a combination of persons and institutions/organizations, that candidates for "B" have multiplied over the years, that the context is recognized to be ecologically and ethically problematic, that the mode has become more and more refined and technologically complicated, and that R cannot be separated from categories like "profit" or "demand and lifestyle," what Douglass Cassel refers to in his chapter as end uses. This would then immediately bring us into the realm of ethical considerations with issues like: division of moral labor between individuals and institutions; questions of responsibility, sustainability, and accountability; analysis of stake holders with their power structures; or moral deliberations of lifestyle and virtues like temperance and justice. This perspective is an integral part of the analysis, not an optional addition.

This analysis would also allow for an analysis of conflicts if we expand it:

> X extracts B from A in C through M because of R, affecting Y.

Y would be stakeholders in the process; we can invite an explicit consideration of violated interests through this version:

> X extracts B from A in C through M because of R, against the interests of Y.

We can also make the potential for conflicts explicit in this version:

> X extracts B from A in C through M because of R, thereby causing damage to Y.

In this form, the analysis would ask for an explicit reflection on those whose lives, interests, properties, or communities have been damaged by mining. This is not an unrealistic appeal to end mining, but a plea for the proper consideration of the full and integral reality of the act. A commitment to IHD would specifically ask to consider in "Y" the most disadvantaged and least privileged stakeholders.

A theological perspective would add specific connections to this analysis, especially in the justifiability of the mode "M," reflection on the reasons "R," and "weighing" of the stakeholders "Y" in the spirit of a "preferential option for the poor." Particular modes "M" of extraction can be severely problematic in and of themselves, such as the use of cyanide to leach gold from ore. The reasons "R" can be economic (generation of profit, state revenue) as well as existential (preservation and expansion of capabilities and a form of life). In any case, the question "who benefits" has to be considered a key aspect in this kind of analysis. That is why the explicit and differentiated consideration of stakeholders "Y" will be indispensable.

Stakeholders of the Essakane mine in Burkina Faso, to name a prominent example, are local communities, regional and national governments, global investors, other mining sites, and global consumers, among others. An integral analysis of the connections between different aspects and different players in the mining industry will lead to contextual ethical reasoning that takes into account the local contexts and dynamics.

Strengthening the theological voice

A theological analysis can add important aspects to an integral analysis of mining. I want to mention three specific connections that can be distilled from *Laudato Si'*. First is the connection between the inner world and the outer world: "The emptier a person's heart is, the more he or she needs things to buy, own and consume" (Francis 2015, §204). This is made even more explicit in a quotation from Pope Benedict (quoted in Francis 2015, §217): "The external deserts in the world are growing, because the internal deserts have become so vast." Second is the connection between the ecological and the social: "The human environment and the natural environment deteriorate together" (§48), so the cry of the earth and the cry of the poor have to be heard at the same time (§49). Finally, there is the connection between knowledge and pain: we must cultivate a "wound of knowledge," daring "to turn what is happening to the world into our own personal suffering" (§19).

Laudato Si' addresses a number of obstacles to IHD, specifically the phenomenon of "rapidification" (§18), the "technocratic paradigm" (§109), "excessive anthropocentrism" (§116; cf. §122), "rampant individualism" (§162), the idea of "maximizing profits" (§109, 195), and the attitude of "self-interested pragmatism" (§215). These reference points provide candidates for further questions for our analysis of the dynamics of "extracting."

Most of these aspects seem to be ethical and not theological in nature. A statement characterizing authentic development could be understood in this way: "Authentic development includes efforts to bring about an integral improvement in the quality of human life" (§147). Categories like "improvement" and "quality of life" seem ethical in nature, and there are undeniably building blocks of an ethics of IHD in the encyclical. But a close reading clearly shows that IHD is a theological and not an ethical concept. We find the strong statement: "A spirituality which forgets God as all-powerful and Creator is not acceptable" (§75). This could be related to the warning that the absence of a sense of mystery is destructive: "When nature is viewed solely as a source of profit and gain, this has serious consequences for society" (§82). The reference to "mystery" is not simply a reference to a category; it is a reference to a spirituality, to an attitude towards life as a whole.

We could read the encyclical as an invitation to a two-fold transformation: (a) the ecological crisis cannot be approached with the means of technological progress; and (b) technological challenges must be transformed into moral concerns. But we cannot stop there. The encyclical invites a further step: moral concerns are translated into spiritual questions (cf. §202). In consistently offering religious language and religious categories, *Laudato Si'* prepares the ground for a particular anchoring of our ecological challenges, for a particular way of framing the stage of the ecological drama.

Dietrich Bonhoeffer pursued a project of translating religious terms into non-religious language in the 1930s and 1940s. The encyclical seems to do the opposite. We see the project of translating non-religious terms in religious language: "nature" becomes "creation" (§76), "land" becomes "a gift from God" (§146), "animals" become "creatures reflecting something of God" (§221), human life becomes a journey "towards the sabbath of eternity" (§243). This religious hermeneutics of the world is part of a sacramental view where "there is a mystical meaning to be found in a leaf, in a mountain trail, in a dewdrop, in a poor person's face" (§233). The roots of the environmental crisis are recognized as "ethical and spiritual" (§9). The theological perspective of *Laudato Si'* is made explicit in the engagement with Patriarch Bartholomew's language of "sin" as it mentions human-made changes to the climate by contaminating the earth's waters, its land, its air, its life (§8). The patriarch is also quoted with a characterization of the world "as a sacrament of communion" (§9), thus inviting a sacramental view of the universe that connects the visible with the invisible, that sees the world as an expression of divine will and love, as "a magnificent book in which God speaks to us" (§12; cf. §85).

A theological perspective accepts the commitment to not leaving the first word and the last word to this world. In fact, the encyclical points out, creation is harmed, according to an address by Pope Benedict XVI, "when we ourselves have the final world" (Francis 2015, §6). And here, one of the fundamental theological statements comes in: "We are not God" (§67).

What is the connection, we could ask, between this theological depth of *Laudato Si'*, mining, and peacebuilding? The answer would be a matter of the level of responses. Conflicts point to deeper issues than a clash of material interests; there are issues like identity, tradition, and honor involved. Conflict resolution cannot be reduced to a technique, but has to be built on trust and reconciliation. The response to the environmental crisis is an ecological conversion, a particular asceticism (§9). A faith perspective can provide motivation to care for nature and the most vulnerable persons (§64), motivations "which make it possible for us to live in harmony, to make sacrifices and to treat others well" (§200). The category of "sacrifice" refers to a fundamental re-ordering of priorities. These are the deep questions that theology can evoke.

Development as depth

The theological (rather than ethical) nature of *Laudato Si'* leads to an understanding of development that cannot be grasped in key terms of moral philosophy like "freedom" or "fairness." These terms are important, but a theological reading of the world and of the relationships within the world touches a dimension that allows and asks for categories such as "mystery," "gift," "meaning," and "service."

Ultimately, the encyclical presents us with a vision of "development as depth." Development, as presented in *Laudato Si'*, is not primarily about progress and living standards; it is not even about maximizing freedoms and quality of life. In one particular passage, Pope Francis introduces the term "depth in life" (§113), which criticizes the superficiality created by the accumulation of constant novelties. This thought presents an alternative to a certain understanding of progress. In fact, the encyclical calls for a change of direction. IHD requires deep change (§60). This change can be linked to a particular understanding of moral and personal growth (§127), growth anchored in deep commitments that may even motivate sacrifices. This is a vision of IHD that is explicit about the cost of "integrating," of accepting vulnerable beings, "however troublesome or inconvenient" (§120).

Here, we move in an area that transcends the language of "quality of life." Let me illustrate the difference between "quality of life" and "depth of life" with an example. Walter Jens was one of the most prominent public intellectuals in Germany. Around 2003, when he was eighty years old, he was diagnosed with dementia. His wife, Inge, accepted caregiving responsibilities and talked about her experience in both the final chapter of her autobiography and in a moving book (Jens 2009; 2016). Jens's words are imbued with melancholy about the loss of a conversation partner, of a "Thou," of the person she loved for decades. Clearly, the quality of life of Inge Jens diminished significantly during those years. But she could not *not* live with her husband, care for him, look after him. This is an expression of a deep commitment that could better be captured by the term, "depth of life."

"Depth of life" points to the identity issues at stake in most conflicts. When rural or indigenous communities face losses from a mining project that is supposedly good for economic development, they may officially gain access to higher "living standards," but they lose identity-conferring aspects of their lives. Andrés McKinley discusses in his chapter how the anti-mining movement, of which the Church in El Salvador was a major leader, used the slogan "Yes to Life, No to Mining." This "Yes to Life" is spoken on a level that is deeper than the material dimension of human existence. It can be framed in spiritual terms. Coming back to conflict resolution we can see that McKinley's emphasis on the commitment of the movement to non-violent action and legislative transformation is based on a spirituality of encountering people and of encountering problems, a spirituality based in a sense of "what really matters" in a local context. Respecting notions of what really matters locally is an indispensable aspect of any peacebuilding effort. Peacebuilding intends to lead to "peace of mind," a spiritual category connected to the idea of what makes life deep and rooted.

Peacebuilding cannot ignore the deep roots of people and communities. Simone Weil, the French philosopher with a deep sense of the spiritual, was asked in 1943 by the Free French Resistance to write a text about rebuilding France (and Europe) after the war. She was asked to reflect on peacebuilding and peaceful rebuilding. In response, Weil (1949) chose to consider not so much structures and institutions, but the needs of the soul. The text, posthumously published under the title *L'Enracinement* (*The Need for Roots*), talks about the need to address the moral and spiritual *malaise* of the time, the need to respond to the dissolution of community by recognizing duties towards humanity. This is the level on which to address the roots of conflicts—on the level of existential roots.

Depth of life is the existential situation of a person who deeply cares about someone or something. This robust concern structures life and gives it weight and profoundness. A theology of IHD will also ask for the integration of suffering, woundedness, vulnerability, limits, and loss. There is a sense of immeasurability and non-functionality. The dignity of the human person is not defined by "functioning well" or by "visible achievements." As *Laudato Si'* expresses, "We forget that the inalienable worth of a human being transcends his or her degree of development" (Francis 2015, §136). The theological vision of development articulated in *Laudato Si'* is a vision of the kind of depth that accepts the cross as a motif of human development. This understanding of depth is built on the foundation of commitments that are love-filled and love-shaped relationships. When we read that "social love is the key to authentic development" (§231), we can see the movement from individual freedoms to social commitments, from personal development to inner growth, from quality of life to depth of life. This is not to say that individual freedoms, personal development, and quality of life are not important. The claim is only that there is a genuine contribution by theology that ethics is not able to offer. Even though there can be an ethical understanding of "social love," the theological meaning cannot be separated from a sense of God's love for creation and God's creatures.

Obviously, this religious sense cannot be presupposed or imposed. But framing mining in terms of "extracting resources from creation" rather than "extracting resources from nature" can provide a helpful perspective for understanding the position of local communities that are, more often than not, rooted in faith traditions. Furthermore, the category "depth of life" is a reminder of a dimension of human existence that cannot be measured, let alone expressed in a language of "risk management," "convenience," or "profit." A theological understanding of development is always at the same time external change and internal transformation. What is being extracted in a mining operation is, then, not only material, but also identity resources. This dynamic has to be taken seriously in the attempt to understand the deeper dimensions of conflicts.

An integral theology of mining and peacebuilding

On May 3, 2019, Pope Francis addressed participants at a meeting on the mining industry. He criticized a particular economic model, "voracious… profit-oriented, shortsighted, and based on the misconception of unlimited economic growth" (n.p.). "Mining for the common good" means mining that it is at the service of the entire human community in the light of the universal destination of goods with a special place of local communities at the table (cf. Francis 2015, §183). Especially vulnerable communities like indigenous communities have to be protected and specifically invited. Mining for the common good also means that it serves the human person with special consideration of human rights. Pope Francis also called attention to a "throwaway culture" and the moral necessity to organize mining operations around a model of a circular economy.

These papal exhortations are framed in theological language, not merely in ethical terms. Francis (2019, n.p.) explicitly refers to a fundamental spiritual attitude: "religious traditions have always presented temperance as a key component of a responsible and ethical lifestyle. Moderation is also vital to save our common home. 'Blessed are the meek, for they shall inherit the earth' (Mt. 5:5)." We leave the context of arguments here and move into the realm of faith narratives which play an important role in the lives of communities. And we reach a context where a spirituality of peacebuilding becomes tangible.

The transformation that *Laudato Si'* calls for and the theological perspective expressed in the encyclical are relevant for the mining industry. Again, there is legitimate space for ethics, as there are many moral challenges with regard to mining: issues with inequality, environmental damage, health risks (Anaf et al. 2019), conflict, and gender issues. The latter include loss of livelihoods through mining, exploitation of women in artisanal mining, sexual exploitation in mining areas, gender effects of mining-induced community changes, and gendered micropolitics of resistance (Adamson 2017; Jenkins 2015; 2017; Macdonald 2018). There is an undeniable need for an "ethics of mining," given the lack of laws and frameworks. There is a need for a global ethics that recognizes "resource interdependence" and the need to consider "international relations of natural resources" (Siegel 2013, 9). There is also a need for the development of legally binding regulations and a role for legislation in creating structures of accountability.

Raymond Offenheiser, in his chapter for this book, discusses the need for "ethics" in the industry; he also points to the necessity of an inclusive dialogue by underlining the need for religious actors to understand the mining industry mindset. This is one side of the matter. The other would be a dialogue that moves beyond ethics and helps agents from the mining industry to understand the concerns and categories of faith-based persons and communities. Understanding the difference between ethics and theology is a first step. Even non-religious persons hold comprehensive ideological commitments and do not stand on neutral ground with regard to the first and last questions asked by theology. The situation we find ourselves in, confronted with the finite nature of our resources, calls for fundamental questions. And this is where I see the place of theology.

Ethics can recommend proper consideration of ecological systems in their sustainability and of communities in their integrity (cf. Carvalho 2017). It can recommend a proper consideration of the human and moral costs of conflicts in mining contexts; it can call for a proper analysis of the losses of vulnerable local communities. Ethics can reflect on the economic macro-conditions and the normative implications of a paradigm of economic adjustment. As part of a "geoethics," an ethics of mining can develop the concept of "responsible mining" (Bice 2016), calling for a proper consideration of stakeholders, a dialogue-centered and community-based approach to the development of sites, appropriate efforts towards environmental protection and ecological sustainability with integrated waste management and energy saving systems, and the provision of a safe and healthy work environment with appropriate remuneration. These elements could ensure that any operation of the extractive industries is embedded in an integral approach that respects the dignity of the person.

There is a lot that an IHD ethics approach to mining can do. But there is also space for a theological approach to mining: *Laudato Si'* mentions mining activities in the context of the pollution of underground water sources (Francis 2015, §29). What would be added to the discourse if the stakeholders of the mining industry were willing to work with categories like "creation," "creature," "sin," and "soul"? Would it make a difference to think about the mining industry in terms of "penultimate life" and the world as having the second and the penultimate word, but not the first and the final? Would it make a difference to turn to the spiritual? *Laudato Si'* states: "There needs to be a distinctive way of looking at things, a way of thinking, policies, an educational programme, a lifestyle and a spirituality which together generate resistance to the assault of the technocratic paradigm" (§111). Francis makes an explicit call to "an ecological spirituality" (§216); this spirituality includes tenderness and compassion (§91), an awe-filled contemplation of creation (§125), a sense of receptivity and gratuity (§237), and a deep sense of beauty (cf. §97, 112). Would it make a difference to introduce the idea of a "spirituality of mining" into the discourse on the extractive industries?

The most concrete imperative that follows from an integral human theology on mining that is committed to an understanding of respect for creation is the rejection of the principle of the maximization of profits (cf. Francis 2015, §195). The idea of maximization points to a human tendency to strive for more. There is no saturation point. There can always be more goods and more profit. With the absence of a saturation point there are two possibilities: external limits (legal restrictions, depletion of resources) or internal limits (a sense of "enough"). The latter requires an attitude towards life as such and towards the world as a whole—in other words, it requires a spirituality. I would like to suggest that a fundamental implication of a theology of IHD approach to mining is the rejection of the idea of a maximization of profits and the acceptance of self-imposed limits. As the Congregation for the Doctrine of the Faith and the Dicastery for Promoting Integral Human Development (2018, §10) point out, "No profit is in fact legitimate when it falls short of the objective of the integral promotion of the human person, the universal destination of goods, and the preferential option for the poor." It seems plausible to make this a standard for justifying mining profits given the industry's massive impact on the planet. The burden of proof moves to those who make profits. And a legalistic answer ("it is not against the law") will not do given the idea of a universal destination of goods and the interests of future generations. This, again, is a spiritual dimension, as it touches upon a fundamental existential position, a way of "being-in-the-world." It points to an understanding of "depth of life."

Given the state of the planet, we need to embrace new beginnings. We need to grow and we need to get rid of the huge armor that we have built, armor that only distances ourselves from the mystery of creation. This cannot happen on the basis of propositions, be they ethical, be they theological: "A commitment this lofty cannot be sustained by doctrine alone, without a spirituality capable of inspiring us" (Francis 2015, §216).

References

Adamson, Rebecca. 2017. "Vulnerability of Women in Extractive Industries." *ANTYAJAA: Indian Journal of Women and Social Change* 2, no. 1: 24–31. 10.1177%2F2455632717714134.

Anaf, Julia, Fran Baum, Matthew Fisher, and Leslie London. 2019. "The Health Impacts of Extractive Industry Transnational Corporations: A Study of Rio Tinto in Australia and Southern Africa." *Globalization and Health* 15, no. 13. 10.1186/s12992-019-0453-2.

Bice, Sara. 2016. *Responsible Mining: Key Principles for Industry Integrity.* London: Routledge. 10.4324/9781315765679.

Bosi, Alfredo. 2012. "Economia et humanismo." *Estudos Avançados* 26, no. 75: 249–266. 10.1590/S0103-40142012000200017.

Carvalho, Fernando P. 2017. "Mining Industry and Sustainable Development: Time for Change." *Food and Energy Security* 6, no. 2: 61–77. 10.1002/fes3.109.

Congregation for the Doctrine of the Faith and the Dicastery for Promoting Integral Human Development. 2018. "'*Oeconomicae et pecuniariae quaestiones.*'

Considerations for an Ethical Discernment Regarding Some Aspects of the Present Economic-Financial System." May 17, 2018. https://press.vatican.va/content/salastampa/en/bollettino/pubblico/2018/05/17/180517a.html.

Cosmao, Vincent. 1970. "Louis-Joseph Lebret, O.P. 1897–1966: From Social Action to the Struggle for Development." *New Blackfriars* 51, no. 597: 62–68. 10.1111/j.1741-2005.1970.tb02029.x.

Francis. 2015. *Laudato Si'*. http://www.vatican.va/content/francesco/en/encyclicals/documents/papa-francesco_20150524_enciclica-laudato-si.html.

Francis. 2019. "Address to Participants at the Meeting Promoted by the Dicastery for Promoting Integral Human Development on the Mining Industry." May 3, 2019. http://www.vatican.va/content/francesco/en/speeches/2019/may/documents/papa-francesco_20190503_incontro-industria-mineraria.html.

Goulet, Denis. 1995. *Development Ethics: A Guide to Theory and Practice*. London: Zed.

James, William. 1985. *Varieties of Religious Experience*. Cambridge, MA: Harvard University Press.

Jenkins, Katy. 2015. "Unearthing Women's Anti-Mining Activism in the Andes: Pachamama and the 'Mad old Women.'" *Antipode* 47, no. 2: 442–460. 10.1111/anti.12126.

Jenkins, Katy. 2017. "Women Anti-Mining Activists' Narratives of Everyday Resistance in the Andes: Staying Put and Carrying on in Peru and Ecuador." *Gender, Place and Culture* 24, no. 10: 1441–1459. 10.1080/0966369X.2017.1387102.

Jens, Inge. 2009. *Unvollständige Erinnerungen*. Reinbek bei Hamburg: Rowohlt.

Jens, Inge. 2016. *Langsames Entschwinden*. Reinbek bei Hamburg: Rowohlt.

Keleher, Lori. 2017. "Toward an Integral Human Development Ethics." *Veritas* 37: 19–34. 10.4067/S0718-92732017000200019.

Keleher, Lori. 2019. "Integral Human Development." In *Routledge Handbook of Development Ethics*, edited by Jay Drydyk and Lori Keleher, 29–34. 10.4324/9781315626796.

Macdonald, Catherine. 2018. "The Role of Gender in the Extractive Industries." In *Extractive Industries: The Management of Resources as a Driver in Sustainable Development*, edited by Tony Addison and Alan Roe, 442–459. Oxford: Oxford University Press. 10.1093/oso/9780198817369.003.0021.

Marcuse, Herbert. 2002 [1964]. *One-Dimensional Man: Studies in the Ideology of Advanced Industrial Society*. New York: Routledge. 10.4324/9780203995211.

Paul VI. 1967. *Populorum Progressio*. https://www.vatican.va/content/paul-vi/en/encyclicals/documents/hf_p-vi_enc_26031967_populorum.html.

Siegel, Shefa. 2013. "The Missing Ethics of Mining." *Ethics & International Affairs* 27, no. 1. https://www.ethicsandinternationalaffairs.org/2013/the-missing-ethics-of-mining-full-text/.

Weil, Simone. 1949. *L'enracinement. Prélude à une déclaration des devoirs envers l'être humain*. Paris: Éditions Gallimard.

Wittgenstein, Ludwig. 1922. *Tractatus Logico-Philosophicus*. London: Kegan Paul.

12 Catholic development ethics, mining, and peace: attending to the market's limitations

Albino Barrera, OP

Introduction: a straightforward link

The link between mining, development ethics, and peace is straightforward. Processes and outcomes in mining are invariably and inevitably shaped by markets, even when occurring in nationalized or socialist, centralized economies. After all, both the input and output of mining are ultimately allocated via markets. Both upstream and downstream requirements of mining pass through markets.

Mining markets, like all markets, have significant limitations. They do not arise nor operate smoothly on their own, but are dependent on institutional preconditions, such as the rule of law and a mechanism for property ownership and enforcement. Furthermore, left on their own, markets privilege allocative efficiency over all other possible social goals. This singular focus on allocative efficiency accounts for the much-touted strength of markets in effectively allocating scarce resources to their most valued uses. There is, to date, no other social institution or arrangement that comes close to what markets are able to orchestrate—coordinating the economic decisions of billions of economic actors and firms widely dispersed across the entire world on what to produce, in what quantities and quality, for whom, when, where, with what kind of inputs, in what manner, and at what price to sell. Unfortunately, such accomplishment often comes at the expense of neglecting or even working against other worthwhile socioeconomic goals, such as equity, social harmony, and sustainability. Herein lies the role of development ethics and the connection to peace.

Development ethics poses a number of critically important questions. To what end is socioeconomic life pursued? Who should benefit and who should bear costs? To what degree are market participants accountable for the consequences of their economic activity? What are their obligations to their fellow market participants? These are but some of the questions raised in this field. Development ethics clearly has a hand in shaping the public economic ethos that sets the formal and informal rules governing socioeconomic life, including market exchange (see Gasper 2012; Goulet 2006). Such public morality determines what market processes and outcomes are

DOI: 10.4324/9781003094272-12

acceptable or unacceptable in the community, and what remedial measures are necessary.[1] These standards change over time and vary across different communities, depending on their respective populations' evolving ethical thinking. Development ethics, in turn, is animated by the philosophical and theological commitments held by the community. It is through development ethics that Catholic social thought and praxis offer their contribution to socioeconomic life, including in the mining sector.

A word of caution: it would be a mistake to juxtapose market economics and development ethics as competing poles with opposed goals. Far from being rivals, market economics and development ethics complement each other. In particular, socioeconomic life would not be sustainable in the long term if it were to operate on the basis of allocative efficiency alone. History and empirical evidence attest to the importance of equity and social harmony in reaching the economic goals of allocative efficiency, growth, and stability. Note that among the most stable and successful communities are those with a sizable middle class (the consequence of equity as a social goal). The six socioeconomic goals—allocative efficiency, growth, stability, equity, social harmony, and sustainability—need each other in order to be complete and enduring in their respective accomplishments. Stated in another way, the economics of market operations call for and require development ethics. Markets need development ethics because while they are adept and incomparable when it comes to effecting allocative efficiency, growth, and stability, they are often deficient when it comes to equity, social harmony, and sustainability.[2] And peace is a necessary condition for those goals.

The market's limitations manifest in different ways across different types of economic activities, including mining. Mining often aggravates the market's limitations, thereby making it that much more important to be attentive to what development ethics has to offer. This essay examines six of the more significant market limitations prevalent in the mining sector and considers what extra-market remedies are needed (development ethics) and what Catholic social thought and praxis have to offer. It is a contribution to what can be called *mining development ethics,* that is, the extra-market interventions and oversight that are needed to put mining sector processes and outcomes in the service of peace and the common good.

Issue #1: Mining's damage to the environment and technical externalities as a market limitation

A clear example of market limitation is the problem of technical externalities, the most famous example of which is pollution. In the absence of government or community regulations, firms can simply dump their waste in the commons (air, water, and land), leaving the rest of the community to suffer the consequences and bear the cost of cleanup. Such a situation is inequitable, threatens peace, and is unsustainable. Even neoclassical economic theory finds it problematic because such technical externalities impede allocative efficiency. The private cost of producing such goods or services is not equal to their social cost. Without constraints, polluting firms can, in effect, free-ride on the rest of the community. Neoclassical economic policy rectifies such deficiencies via taxes, outright regulation, or a combination (see Pigou 1920 [2013]). This is one issue in which development ethics and neoclassical economic theory converge in their conclusion on the need for extra-market intervention to rectify technical externalities.

Among the most common complaints against mining industries is their damage to the environment. Without regulatory oversight, mining companies tend to leave contaminated air and water, toxic dumps, tailings, and badly scarred landscapes in the wake of their operations.[3] Both neoclassical economic theory and development ethics are in agreement that such conduct is unacceptable and has to be rectified. Reaching a broad societal consensus on the need to hold mining companies accountable for ecological damage should be relatively easy because there are strong economic and ethical reasons behind such clamor. Mining firms and their shareholders should not be able to reap the benefits while leaving the associated costs to unwitting third parties (see Okonta and Douglas, 2001; Martinez-Alier 2001; McBeth 2018). Allowing them to leave such ecological damage unaddressed has clear adverse impacts on the common good, and conflict and instability are exacerbated when there is inequity in the distribution of costs and benefits.

Ecological protection is a position of mining development ethics that enjoys wide agreement. Catholic social thought and praxis offer their principle of stewardship, that is, the moral obligation to care for and use the gifts of God, the earth and its goods, wisely and well. In *Laudato Si'*, Pope Francis (2015) considered stewardship of the earth through the concept of integral ecology, which calls for integrating environmental concerns with other aspects of social and economic justice, such as justice for the poor, who often suffer disproportionately from the environmental harm caused by mining and the way mining's environmental degradation can drive conflict.

Issue #2: Mining and the amoral nature of markets

Markets facilitate economic exchange but are not designed to evaluate, much less judge, the rightness or wrongness of such transactions. Markets only require that one party has purchasing power and another has possession of a good or service in order to consummate an exchange. Unfortunately, this sidesteps the fundamental question of how the community ought to allocate the scarce goods of the earth. By effort, allocative efficiency, merit, need, contribution, parity, or another standard? There are as many criteria as there are philosophical commitments and views on what the good life entails.

Markets are not designed to make such choices for the larger community because they are merely contractual social devices that we use to orchestrate a whole series of difficult decisions on what to produce, for whom, how much, in what manner, and who gets priority. They do not distinguish between genuine needs and whimsical wants because markets consider only purchasing and bargaining power in consummating exchanges. This gives preeminence to consumer sovereignty. Allocative efficiency, in effect, trumps all the other possible and perhaps even more important goals of socio-economic life.

This is an unworkable market dynamic whether viewed from a practical angle or from the viewpoint of ethics. In actual practice, markets can only function within the framework of a minimum number of extra-market interventions and institutions (such as rule of law, private property ownership and enforcement, etc.). Thus, government and civil society have significant roles in the marketplace. In their own turn, government and civil society are shaped by the philosophical and religious commitments of their membership and leadership. This gives rise to the manifold and varied extra-market interventions of different types of political economies (e.g., Western market capitalism, China's state capitalism, Cuba and Venezuela's socialist economies). Development ethics and Catholic social thought have much to contribute in this societal process of reining in market processes and supplementing their underlying singular pursuit of allocative efficiency. For example, consider some of the problematic outcomes in the mining sector in the wake of an exclusive reliance on purchasing and bargaining power in allocating societal resources: unsafe working conditions, one-sided revenue sharing between a multinational corporation (MNC) and a host country, inequitable revenue sharing within the host country, and inadvertent material cooperation in wrongdoing. We examine each of these briefly.

Unsafe and poor working conditions

Because markets are allocative rather than normative in their function, they do not and cannot deal with coercive contracts or unequal bargaining power and resulting economic compulsion. The mining industry provides the earliest example of this market limitation in the modern economy. Much has been written on working conditions in British mines early in the Industrial Revolution, and we need not rehearse them here. It is sufficient to note that such abuses precipitated the earliest examples of widespread social unrest that ultimately compelled extra-market intervention, when Parliament passed legislation that mandated workplace safeguards (Mills 2010).

Today, we see the same dynamic at work. The poor and the unemployed accept unsafe and physically taxing and unhealthy working conditions in mines because they are desperate for income. Social legislation, potential lawsuits, scrutiny from non-governmental organizations, or the risk of bad publicity level the playing field and compensate for workers' inability to negotiate better conditions for themselves. The need for such extra-market interventions arises because markets consummate economic exchanges solely on the basis of purchasing and bargaining power, not the quality—much less the equity—of such transactions.

Catholic social thought and praxis have long championed protection for workers given their vulnerability to abuse. The seminal social encyclical *Rerum Novarum* was a response to the abuses of the newly inaugurated industrial age of the nineteenth century (Leo XIII 1891). This was followed ninety years later by a commemorative encyclical, *Laborem Exercens*, that affirmed once again the importance of providing living wages and humane working conditions (John Paul II 1981). Its principle of the primacy of labor reminds business decision-makers that workers cannot be viewed merely as factors of production no different from capital or raw materials. Rather, they must be regarded for what they are—human beings—and must be treated accordingly, along with their dependents. Catholic social teaching has much to offer regarding widespread concern over poor or unsafe working conditions and pay in the mining sector. This is a central point of what could be called Catholic mining development ethics.

One-sided revenue sharing between companies and host countries

A second problem that arises in the mining sector is the lopsided nature of revenue sharing with the host country, stemming from the market's sole reliance on purchasing and bargaining power in consummating exchanges. Countries in the Global South are often compelled to enter into disadvantageous contracts with mining MNCs.

Empirical evidence from economic history tells us that most countries' economic center of gravity moves from agriculture to industry to services as they develop (Herrendorf, Rogerson, and Valentinyi 2013). Countries in the initial stages of their economic development rely heavily on their agricultural sector and mineral resources. Pre-Industrial Revolution United Kingdom output and pre-nineteenth century United States GDP were comprised predominantly of agriculture and mining. It is the same pattern that we see today. Contemporary Global South countries are heavily reliant on these sectors.

Such countries need to earn enormous sums of foreign exchange in order to pay for their imports of vital technology, raw materials, and capital goods. Mineral resources are a ready source of such foreign exchange, given the global demand for such commodities. Unfortunately, countries in the Global South often do not have the necessary capacity to extract, process, and sell these mineral resources themselves in global markets. They also do not have the funds to invest in what is often an extremely expensive and capital-intensive effort. They are thus heavily reliant on MNCs. This can lead to lopsided revenue-sharing contracts between MNCs and host nations. Extra-market safeguards are needed to head off such uneven contracts. These include NGO advocacy, technical advice from multilateral agencies such as the World Bank and the United Nations, and international pressure against exploitative practices (see Mann 2015). Nevertheless, despite these extra-market interventions, we still have many disputes and allegations over exploitative revenue sharing arrangements between MNCs and host countries (see Burianski, Nyer, and Kuhnle 2019; Kimani 2009).

This is yet another illustration of a place where development ethics and Catholic social thought and praxis can offer needed extra-market norms in the mining sector. For example, MNCs should not view and treat host countries and local communities in a purely transactional manner, giving to the latter only that which is specified by their contracts or what is permissible under market operations. The principles of solidarity, universal destination of goods, subsidiarity, and integral human development call on these MNCs to treat their host countries and local communities as their equals. One-sided contracts are unacceptable because they are in effect taking advantage of the plight, ignorance, or weakness of one's trading partner. This is exploitation. The universal destination of goods is an affirmation that the earth is meant to benefit all, regardless of how titles of ownership are assigned. Integral human development should remind MNCs that authentic development is a joint effort that can only be completed together. The principle of subsidiarity should make them sensitive to their obligations to use their advantages in technical know-how, knowledge of markets, and vast experience in mining to good effect in helping poor nations and local communities develop and benefit from their natural resources. They are unable to do so on their own for want of the technical skills, capital, and market networks—the very strengths of the MNCs. Indeed, Catholic social thought can go a long way in making MNCs see that the mining industry is an ideal venue not only to maximize returns for their shareholders, but more importantly, to make a difference in the lives of people. Unlike international arbitrators and multilateral agencies, Catholic social thought cannot legally compel MNCs, but they can do something even better—appeal to their better selves via reason.

Mining and the sharing of revenues within the host country

Markets work toward allocative efficiency, not equity. Markets merely facilitate economic exchanges and are not concerned with their distributional consequences. Markets do not second-guess the equity of these economic transactions because they do not limit inequality. In fact, the logic of survival of the fittest can turn it into a form of buccaneer capitalism—a winner-take-all economy. Thus, it is entirely possible for the gains and benefits produced by markets to be concentrated in the hands of a few within the country (Piketty 2017).

Nigeria is an example of the failure of oil wealth to lift the general population out of poverty (Slav 2018). In contrast, note the example of Norway and its disposition of its share of North Sea oil proceeds (Norges Bank n.d.). These differing examples underscore the importance of extra-market oversight and intervention (e.g., Norwegian public policy) in guiding market processes and in shaping market outcomes.

The principles of solidarity and relative equality from Catholic social thought and praxis are useful tools for development ethics in attending to these limitations in the extractive sector. The principle of solidarity calls for a genuine and active concern for the well-being of others because we see in them ourselves and a fellow child of God. The neighbor is viewed and treated as an equal. It is unacceptable to have pockets of great wealth and power amidst a vast sea of poverty because it reflects indifference to the plight of the poor and creates conditions that contribute to violent conflict. Moreover, relative equality is a necessary condition to achieving harmony in any community. Beyond a certain threshold, inequalities in wealth, power, and lifestyles not only breed dangerous social unrest, but they also mean that substantial segments of the population do not have the necessary resources to participate meaningfully in the common socioeconomic life. Thus, inequalities can only be permitted to go so far in any functional community.

Inadvertent material cooperation in wrongdoing

As mentioned earlier, markets are merely social institutions that allocate burdens and benefits across the community based on purchasing and bargaining power. On their own, they are not designed to make normative judgements on the resulting market outcomes. Thus, it is fairly common for market participants to unwittingly enable wrongdoing somewhere in the community. After all, market decisions have extensive ripple effects, both beneficial (positive externalities) and adverse (disexternalities). Market participants' resulting material cooperation in wrongdoing down the line can either be blameworthy or blameless, depending on whether or not they could have taken preemptive or remedial action to prevent or mitigate the subsequent wrongdoing.

For example, the enormous wealth that the mining sector produces creates numerous problems, such as engendering corruption and providing revenues for the waging of war or violence. Hence, we have the campaign against "blood diamonds" and the push to identify the source of coltan in the manufacture of cell phones (Campbell 2012; Nest 2011). The chapter by Rigobert Minani in this book explains how coltan funded war in the Democratic Republic of the Congo. Besides fueling such conflicts, mining can also unwittingly enable corrupt or failing governments, as was also the case in the Democratic Republic of the Congo, with corruption funneling mining revenues into the pockets of officials in the Kabila regime. Sierra

Leone's vast diamond resources have been both a blessing and a bane, engendering frenzied theft and corruption, not to mention the thousands of lives lost from the wars that such diamonds funded and sustained.

The helplessness of markets in the face of these harmful ripple effects shows the need for development ethics in preventing or at least minimizing these unintended consequences from mining's huge revenue streams. There are many historical instances of extra-market interventions to prevent markets from enabling wrongdoing, such as the Kimberly Process to certify diamonds being conflict-free, or the United States' Dodd-Frank Act that required US companies to not source minerals connected to violent conflict.

To summarize the elements of this issue, markets allocate according to purchasing and bargaining power, with allocative efficiency as the primary goal that trumps all other possible goals of the economy. Concerns in mining development ethics over poor working conditions, economic compulsion, one-sided revenue sharing contracts, and great inequality are not unique to Catholic thought and praxis. There is widespread agreement on the need to address these market deficiencies, but Catholic moral theology and Catholic social teaching lend useful resources and concepts for analyzing these situations. This includes the idea of material cooperation with evil, which can result either from wrongdoing or from failure to prevent wrongdoing when the capacity to do so is present.

Issue #3: Mining and markets' lack of social consciousness

Markets do not go beyond contractual obligations because they are not designed for other types of obligations, such as moral obligations. One ethical problem that arises is the lack of social consciousness.

Scholars and policymakers have flagged the problem of extractive-export enclaves with little forward or backward linkage to the rest of the country (see Phelps, Atienza, and Arias 2015). Backward linkage pertains to the supply chain supporting the extractive operations, such as the stream of engineers, technicians, capital equipment, fuel, raw materials, and sundry tools vital for the industry's operations. Extractive-export enclaves often import these from overseas and can avoid sourcing their requirements domestically. As a result, the local economy and the local community do not share in the economic value created by mining.

Forward linkages pertain to the further processing of extracted resources into their various intermediate or final products. Examples of such forward linkage includes the processing of extracted ores within the host country to produce raw metals that are in turn processed even further into end products such as steel sheets, wires, etc. Forward linkages are highly sought because of the jobs and industries they create within the host country. Clearly, it is in the interest of the host country to have vibrant backward and forward linkages with their mining sector, especially if these host countries are just starting on their development path with little else but their mineral wealth. Not only do the local citizens earn a livelihood, but just as importantly, they gain skills.

Unfortunately, extractive-export enclaves arise with little backward or forward linkages with the rest of the host economy. Consequently, we see pockets of great wealth and advanced development in the immediate region where extractive operations are located while the rest of the country languishes in poverty. Left on their own, markets will not correct such imbalance. In fact, economic theory itself may further support and strengthen such enclaves if it can be empirically shown that the host country has a comparative advantage in the extraction and export of such mineral resources and not in the further processing of such minerals due to the lack, for example, of capital and cheap energy sources.

Left on their own, markets are concerned only with fulfilling contractual obligations. Mining firms satisfy their obligations once they fulfill their contractual obligations to the host country or local community. The economics of markets is such that economic agents are under no obligation to further anything else beyond their own interests. Markets are not geared for enforcing or living up to moral obligations. Corporate social responsibility is an extra-market intervention or mechanism that pushes market participants to go beyond their own interests and think of the welfare of the larger community of which they are a part. In recent years, as discussed in the chapter by Douglass Cassel, there has been a push to develop stronger and more legally binding standards for business and human rights that are not as easily manipulated by corporations who can sometimes simply throw money at a community building project and think they have satisfied their obligations. But "corporate social responsibility" can often be practiced without real social consciousness, rather than with the more robust understanding that development ethicists or business and human rights advocates might put forward.

Development ethics raises the issue of the moral obligations of mining firms to help their host communities and countries along the path of development by giving them opportunities to move up the value chain or increase their value-added contribution to the local processing of the extracted minerals. Related to this, of course, is the danger of the opposite extreme of paternalism, of being overly directive in telling host nations how they ought to traverse their path to development. Of course, it is possible for local populations to choose not to move up the value chain and to be content with enjoying their share of the revenues from extractive-export enclaves. Thus, we have literature on "the natural-resource curse," in which entire populations become complacent, develop few skills or a work ethic, and end up worse off than their resource-poor neighbors (see Humphreys, Sachs, and Stiglitz 2007; Frankel 2010).

From a purely economic-contractual point of view, there are no economic, much less moral, obligations beyond what is stipulated in contracts. Not so for development ethics, and not so especially for Catholic social thought and praxis, for the following reasons. First, integral human development is about (1) the development of the whole person—body, mind, and spirit—and (2) the development of every person. This second requirement of integral human development means that it is a journey that can only be completed together with our neighbor. We cannot claim to have reached our own integral human development if our neighbor, whom we could readily help, wallows in unattended need. Clemens Sedmak's chapter explains how these tenets of integral human development connect to peacebuilding. They amount to a model of development that prioritizes conditions in which conflicts can be handled in ways that preserve co-being rather than having some actors, like mining corporations with huge political, economic, and legal advantages, trying to gain maximum benefits.

Second, we have two parts of the principle of subsidiarity. The first part of this principle notes that higher bodies should not arrogate for themselves functions that lower bodies or individuals are able to perform on their own. Private initiative is to be fostered. The second part of this principle balances the first part by observing that it is the obligation of higher bodies or individuals to step in and provide assistance once lower bodies or individuals are in trouble and no longer able to function on behalf of the common good (Barrera 2001, 263–65). By the second part of the principle of subsidiarity, mining companies are bound by moral obligations to assist in the promotion of the welfare of their host communities, and to assist with problems that require higher levels of coordination and resources, such as security or integrating local operations with national initiatives like peace processes. Note that these responsibilities apply not only to governments but to any entity that has the capacity and resources to address the community's unmet needs.

The unsatisfactory nature of extractive-export enclaves is an issue where both economics and Catholic social tradition converge in their respective positions. It is also an illustration of where extra-market intervention is needed because allocative efficiency itself may dictate that the country remain a commodity exporter, while higher-value processing is reserved for other countries that have the comparative advantage.

Both the principles of subsidiarity and integral human development are also relevant for the earlier issue of one-sided or coercive contracts (issue no. 2 above). To take advantage of people in their need or in their ignorance is exploitation. Such predatory conduct, however, is not limited to coercive or one-sided contracts alone. We can be just as predatory with a failure by omission when we fail to prevent harm when we could easily do so. Mining firms that can do more to further the well-being of their host communities ought to do so. The preferential option for the poor makes this obligation even more compelling in the case of impoverished host communities. Economic exchange has to be of mutual benefit to all parties. What constitutes mutual benefit is where development ethics and Catholic social tradition can make a difference. For Catholic social tradition, it is nothing less than integral human development. Moreover, for development ethics and Catholic social tradition, outreach to the local community beyond what market contracts stipulate is not supererogatory but is a moral obligation. The poorer and less secure the host communities, the stronger and the more extensive are the moral obligations.

Issue #4: Mining and market limitations due to ownership externalities

The problem of ownership externalities has long been known in the economic literature (see Bator 1958, 351-79). Markets often fail to assign an accurate exchange price for goods and services because they are unable to properly account for all inputs. Economic agents are unable to see or are not liable for the full accounting of costs and benefits. Hence, they do not sufficiently value the benefits they are enjoying from "the economic commons." Market participants do not see that they have a "social mortgage" in being able to so readily and easily participate in market exchange. This is so because there are many costs and inputs to maintaining markets that are not accurately priced, if they are even included at all. For example, take the price of rare earth metals that are essential for electronics. It does not include previous decades' investments in developing the basic science and technical know-how that enables users to harness the unique properties of these minerals. Or, consider the global banking infrastructure and international arrangements that facilitate international trade. These institutions and formal and informal rules took centuries to build, often at great expense. These underlying, unseen costs and inputs are often taken for granted and are not priced into market exchange. Such accounting problems are called ownership externalities.

Ownership externalities are another reason that mining firms cannot limit themselves only to their contractual obligations to their host communities. These mining firms are drawing benefits from their host communities—benefits that would most likely not have been accurately priced into the revenue-sharing contract. For example, many of these host nations have had a long history of strife and nation-building. Costs were incurred over time in building up the social, human, and physical infrastructure of the host communities. These costs are often not fully reflected in the price of participating in the marketplace. Market participants are thus able to draw benefits from the marketplace without having to share in the cost of many vital inputs that make it possible for such local markets to exist. Thus, it is very likely that mining firms avail themselves of many benefits from their host communities that have not been properly priced by the marketplace. This is especially likely in the Global South where markets and government fiscal structures are still rudimentary.

This is yet another area where the contributions of Catholic social thought and praxis stand out. Ownership externalities prevent us from establishing accurately what is due to others in the marketplace. This is a significant deficiency, because justice is about giving others their due. Catholic social tradition offers a solution to this dilemma because it sees social life as operating not merely on the basis of justice. In fact, justice by itself is insufficient. Justice must find its perfection in charity (see Aquinas 1911–1925, II-II, q. 23, art. 6, 8). Thus, in situations where what is due to others cannot be ascertained, it is incumbent on market participants to act on the basis of charity itself. Charity, by its nature, is unmeasured and entails self-giving. In practical terms, this means that in the event of uncertainty or difficulty in establishing what is due to others, it is better to be self-sacrificial and to give more according to one's means. For example, it is difficult to establish accurately our personal contribution to global warming because of our consumption or travel habits. Moreover, we are not obligated by law to compensate for such damage. Nevertheless, some people take it upon themselves to pay for voluntary carbon offsets or to curtail their own consumption as a matter of civic duty, principle, or conscience. In the sensibility of Catholic social tradition, they are motivated by charity. In the case of mining companies, they ought to do more for their host communities beyond what their contracts stipulate, not to promote their public image at home and abroad, but out of a genuine concern for the well-being of their host communities. Charity, in this sense, is to give without thought of return, thinking only of the good of the recipient.

In sum, the phenomenon of ownership externalities and subsequent mispricing in the marketplace are further arguments for why mining firms, or any other firms for that matter, have obligations over and above those stipulated by their contracts. Accounting difficulties in keeping track of inputs and the true social and private cost of goods and services they draw from the community impose these non-contractual obligations.

Issue #5: Mining and the market's inability to establish intrinsic or non-economic value

The preceding issue on ownership externalities arises because of the technical limitations of accounting. Market prices are inaccurate because they do not include many unmeasured or incommensurable costs and benefits. But even if we assume, for the sake of argument, that we are able to arrive at accurate market prices, there is an even deeper problem: our inability to arrive at the true value of certain goods or services.

There is a difference between price and value. Price pertains to the quantity of currency to which buyers and sellers agree in order to consummate trade. Price is essential in order to effect economic exchange. Price is determined only by consumer sovereignty, that is, what people are willing to pay for such goods or services. In contrast, value pertains to the *intrinsic worth* of goods and services.

This difference between price and value can be easily understood by recalling the water-diamond paradox of introductory economics classes. Water is clearly of much greater value compared to diamonds because it is essential for life. If so, why is the market price of diamonds manyfold that of water? The response is, of course, that market price is determined by scarcity. Given its greater abundance relative to diamonds, the market price of water is based on what consumers are willing to pay for the last use for which water can be employed, such as watering the lawn, flushing the toilet, or washing the car. Another example is the compensation of teachers and nurses. These professionals provide services that are vital for life and the well-being of the community. But their salaries, or market price, are far lower than their true value, that is, their intrinsic contribution to the quality of life.

A good scriptural illustration of the difference between exchange price, on the one hand, and intrinsic value, on the other hand, is the Jubilee Law (Lev. 25). The market price of land, when it changes hands, is the sum of the market price of the harvests that can be reaped from that land before it is returned in the next Jubilee year. In contrast, the intrinsic value of land is incalculable, or one could say, non-existent, because land belongs to God (Lev. 25:23), and humans are merely tenants. Thus, land cannot be sold in perpetuity, and only according to the market price of its harvests.

The market's limitation in determining the intrinsic value of goods and services is particularly worrisome in the case of mining, which involves the depletion of non-renewable resources. Economists have long been dissatisfied with the way natural resources are priced in the marketplace (see Nordhaus and Tobin 1973, 509–64). Take the case of the calculation of GDP, the sum total of all the goods and services produced within the borders of the nation. The natural resources that are extracted in a particular time period are incorporated into the calculation of GDP at the market price at which they were sold. Such pricing, however, is inaccurate, especially in determining the wealth of the nation, because the extraction, export, and use of such minerals means that there is less left for future generations of citizens. There is no intrinsic value ascribed to natural resources in the ground. They are priced only according to what they will command in the marketplace.

Similarly, the cost of the ecological damage that mining industries inflict is not properly priced or valued in the marketplace, as in the case of global warming caused by the burning of fossil fuels or the flaring of waste gas. Or, take the case of the enormous amounts of fresh water that are used in mining. The market price of metals and minerals does not incorporate the value of depleted or polluted water, but only what consumers are willing for pay for the mined products. The fresh water supplies used or damaged are not valued for their intrinsic worth, especially for future generations.

Development ethics has much work to do vis-à-vis this market deficiency. There is a need to change contemporary behavior and attitudes regarding the goods of the earth. Market participants should think in terms not only of the exchange price of these goods, but also of their intrinsic value. The calculation of a nation's GDP or of global wealth or income has to factor in the depletion of non-renewable resources. Incorporating intrinsic value into exchange value is relevant for two important issues: (1) revenue sharing with the host country/community and (2) intertemporal equity.

Host communities are clearly at a disadvantage if they are paid for their natural resources based only on exchange price, without taking into account the intrinsic value of these resources. Thus, host nations/communities are underpaid. Such misvaluation means that economic growth from mining is overstated, because the damage to the environment and the depletion of nonrenewable resources are insufficiently considered. This also means that it becomes even more difficult to attain or approximate contemporary allocative efficiency given the mispricing of factor inputs. Mispricing, in this case, is due to the mis-valuation of underlying goods and services themselves, instead of accounting limitations that give rise to the ownership externalities in the preceding issue.

Mis-valuation is central to the issue of intertemporal equity. Non-renewable resources are undervalued, and as a result, they are misused or over-used. Consequently, there is a much faster rate of exhaustion of these supplies of non-renewable resources than is warranted. There will be that much less for the next generations as part of their patrimony from contemporary generations.

Mis-valuation is relevant not only for intergenerational equity but also for intertemporal allocative efficiency. One justification for the use of non-renewable resources by earlier generations was that they are able to expand technological know-how and improve the stock of capital for the benefit of succeeding generations. In other words, in exchange for a smaller supply of non-renewable resources, future generations will be able to enjoy a much more advanced technological base and be equipped with more capital. There is a trade-off between conserving non-renewable resources for the sake of future generations and the amount of know-how and capital that the current generation is able to bequeath them. There is need to find a balance between pushing the envelope of contemporary technologies, on the one hand, and keeping natural resources undeveloped and untapped, on the other hand, in order to conserve them.

Solving for intertemporal allocative efficiency is a much more complex process than that of finding out what contemporary allocative efficiency requires. We compound this difficulty by using the wrong valuation of these non-renewable resources because we are using only their exchange price and have not incorporated their intrinsic value. The major difficulty, of course, is in establishing the intrinsic value of various resources. This is an area of possible collaboration between the fields of development ethics, public policy, and economics.

Imputing intrinsic value is a technical field. Nevertheless, Catholic social thought still has an important contribution to make: a rationale for imputing intrinsic value to the goods of the earth over and above their exchange value. The goods of the earth are a gift from God meant to benefit all, both in the present and in the future. This principle of the universal destination of goods is vividly affirmed in the creation accounts in Genesis. Humanity is entrusted with the mandate to exercise stewardship over God's creation, even as we draw from the earth whatever it is we need for daily living.

A final word of caution: imputing value to mineral resources based on their intrinsic worth will increase current prices of these commodities. Such price increases will in effect be a de facto redistribution of income and wealth. For example, people with a large portion of their budgets devoted to the purchase and use of mineral resources will find their real incomes shrink. In other words, mining development ethics and Catholic social thought would require sacrifice on the part of current market participants. Of course, this is not the first time that people have been invited to willingly sacrifice their current consumption, as in the case of carbon taxes and voluntary carbon offsets. The principle of solidarity is important in this regard—solidarity with brothers and sisters yet unborn.

Issue #6: Mining and the question of ownership and usufruct of the goods of the earth

I have reserved for last the most intractable problem, but the most fundamental one that must be addressed. Markets merely accept, and do not second guess, the ownership-usufruct framework set by the extra-market legal institutions prevailing in the communities in which they operate. In contrast, nations, political economies, development ethics, and indeed the community of nations must address or constantly revisit fundamental questions, such as: who owns the goods of the earth, who should have usufruct rights over them, and how should the costs and benefits from these resources be distributed across the community? These bedrock issues determine how extractive activities are conducted and how their costs and benefits are distributed.

In the case of mining, there is a longstanding de facto global rule that nations exercise ownership and usufruct rights over natural resources within their borders. The 1982 UN Convention on the Law of the Sea sets a two-hundred-mile exclusion zone offshore for purposes of delineating the boundaries of nations' economic claims (art. 57; see also Hollick 1977). Governments have pursued different paths to making use of these natural resources. Some have nationalized extractive industries in key natural resources, such as oil. Others rely on a more market-based approach of letting private firms extract and sell these natural resources in exchange for taxes, lease payments, and revenue sharing, among many other methods.

These different approaches have significant ramifications for who bears the cost and who reaps the benefits. For example, nationalized extractive industries better benefit the general population (at least in theory), while a more market-oriented approach allows entrepreneurs who undertake the risks of development and investment to reap most of the benefits, as in the case of the United States and other Global North nations.

Of concern for us is the question of which method is appropriate. The criteria for making this assessment vary depending on the philosophical commitments of evaluators. Neoclassical economics would most likely privilege allocative efficiency as the measure to use, while development ethics and Catholic social tradition would most likely use a combination of all six socioeconomic goals in varying degrees, again depending on the priorities of the evaluators. Despite this difficulty and the lack of consensus on an ideal or common standard for evaluation, the question of which method is appropriate for mining has to be addressed head on because it directly impinges on the distribution of the burdens and benefits of the metals and minerals, for contemporary generations but also, of equal import, for future ones.

The question of which approach is appropriate points us to even more fundamental questions that we must constantly revisit: (1) Where do we draw the line of ownership and usufructuary rights: at the local, provincial, national, regional, or global level, or a combination of these, and (2) on what bases do we do so? Who exercises rightful ownership and usufructuary claims over these resources? For example, who ought to have ownership-usufructuary rights to the shale basin in the North American continent? The local communities, the state, or the federal government? The main contribution of Catholic social thought in this regard is its principle of the universal destination of goods. Should Catholic social tradition advocate for the global patrimony of all key natural resources in light of its universal access principle (Pontifical Council for Justice and Peace 2004, §171–184) and its fundamental premise that the goods of the earth are a gift for all from God, meant to benefit all? Where does one draw the line?

These fundamental questions have ramifications for all the other issues and questions downstream. For example, do national governments have the right to assign property or usufructuary rights over natural resources to private groups? Do nations have the right to nationalize and benefit exclusively from their resource wealth, or should there be a global fund that distributes these resources worldwide in view of the central importance of metals and minerals in the modern industrial economy and given the universal access principle? Can nations use extractives as instruments of foreign policy, as in the case of the two oil embargoes of the 1970s? Given their vital role in the digital age, should rare earth minerals be considered a global patrimony rather than be controlled by the nations in which they are found?[4]

Even if we were to make the heroic assumption of a global acceptance of the universal access principle, we nevertheless have to balance this with the pragmatic reality of having to avoid the tragedy of the commons and the very real problem of free-ridership. There will still be a need to provide incentives for risk-taking. Work effort and initiative will have to be properly compensated. Nevertheless, satisfying these requirements simultaneously requires much deliberation. We have to find a balance between the Lockean claim to property ownership[5] and the Lockean proviso.[6]

In sum, the questions of who owns and who has usufructuary rights over the goods of the earth and on what bases must be addressed because they determine the distribution of burdens and benefits from the mining sector. They are relevant in arriving at the appropriate arrangements in extracting natural resources, such as whether or not such activities should be nationalized or privatized. Markets simply operate within the formal and informal norms set by the community. Development ethics has a hand in shaping these formal and informal rules. In the absence of extra-market norms, market praxis becomes the norm by default based on purchasing power and existing ownership claims, no matter how unsatisfactory or damaging they might be. The absence of such norms can invite conflict over ownership and prevent mining operations from being harmonized with other objectives, like brokering or implementing peace agreements or ensuring environmental protection.

Conclusion

Development ethics is not an optional add-on to development economics or market operations. Rather, it is necessary given the significant limitations of market processes and outcomes. Extra-market oversight and intervention are needed because markets, on their own, will not attend to the community's well-being, only to the allocative efficiency of its scarce resources. In fact, the whole field of development ethics is about the need for extra-market intervention because left on their own, unfettered market operations are unable to reach larger societal goals, like implementing peace or advancing integral human development. Extra-market institutions are a necessary condition for markets to operate smoothly and sustainably. They are needed if markets are to exist at all, and we see this phenomenon in the mining sector.

Some issues in mining development ethics are fairly straightforward and uncontroversial for most reasonable people, such as the need to rectify pollution and damage to the environment, the need to improve safety and working conditions, the inappropriate use of enormous revenues from natural resources for war and civil strife, the skewed revenue-sharing from one-sided contracts, the inequality in the distribution of revenues within the host nations and local communities, and the danger of fostering corruption in the host communities. Other issues are gray areas, such as intergenerational equity, the problem of distant harms and where the limits of responsibility are, and the obligation of mining sectors not to be isolated export enclaves but to help their host nations to move up the value chain of economic activity. There are also the questions of the proper valuation of the goods of the earth and of who should exercise ownership-usufructuary rights over the goods of the earth.

What is distinctive with Catholic social thought and praxis vis-à-vis these issues and extractive development ethics? First, it has a holistic understanding and practice of development. Another name for peace is development (Paul VI 1967, §76–77). Development is a necessary condition for peace. However, it is not a sufficient condition. There is need to strike a balance between two extremes. On the one hand, there is the danger of an otherworldly dismissive attitude that views socioeconomic life as mundane, secondary, or unimportant. There is a necessary economic dimension to human flourishing. Getting our economics right is important as part of human nature since we are corporeal and social. For example, even as we acknowledge the need to impute an intrinsic value to mineral resources beyond their exchange value, we must be realistic as well in recognizing that arriving at such a value is extremely difficult. Even more difficult is implementing it in practice, especially in light of the much higher prices that will lead contemporary generations to curtail their consumption.

On the other hand, there is also need to avoid the other extreme of economism, that is, of privileging economic factors above all else. From this viewpoint, development is primarily an economic phenomenon and is measured by economic growth. In contrast, Catholic social tradition is adamant that economics is *not* an end in itself but is merely instrumental. It is at the service of integral human development. As we have seen above, there is a two-fold nature to integral human development: (1) the development of the whole person (which explains why development is not merely nor primarily economic) and (2) the development of every person (which explains the importance of distribution and mutual benefits in economic exchange). Catholic social tradition proposes a much broader definition of development and imbues it with much larger goals. Our neighbor's development is our development too, and vice versa. We are necessary conditions to each other's development. This has enormous ramifications for mining firms' obligations to all their stakeholders. It means that their relationship with their stakeholders is not merely contractual, as the markets would have it, but is familial in nature, especially for their workers and their host communities. They are invited to act accordingly.

Another risk of economism is assuming that economic growth is a direct line to peace. Violent conflict often involves many factors that are not merely economic, such as historically rooted animosities, political power, or competing claims about the value of land being mostly about its resources versus being about ancestral claims for indigenous heritage. Assuming economic growth can in and of itself solve challenges of peace ignores other needed elements and processes. And at times, development ethics might call for allocative efficiency and market growth in the mining sector to stand aside to allow other priorities to supersede them.

A second distinctive Catholic social tradition contribution is that, as new creatures in Christ, humans ought to act according to who they have become. Thus, for Christians working in the mining sector, good works should flow from their faith. Mining development ethics should become integral in their conduct—and not merely for post-hoc damage control, but intrinsic right from the very start. Mining ethics is not deontological, with a list of dos and don'ts, but is teleological, because it pushes people to better self-understanding and considers the vocation of all Christians to strive for peace and the common good.

Third, the Catholic social tradition holds that creation is a gift from God meant to benefit all. Such a conviction creates two claims for mining development ethics. First, all are accountable to the Creator. The proper attitude to take, vis-à-vis the goods of the earth for all mining firms, their shareholders, and national governments, is not a proprietary posture, but one of stewardship. There are limits to what we can do with the goods of the earth. Second, there is the moral obligation to ensure that all are able to benefit from the goods of the earth regardless of how titles of ownership are assigned.

Notes

1 For example, most communities use social safety nets to define a "moral floor" beneath which people will not be allowed to sink. Western capitalism give preeminence to individual liberties and the free pursuit of private initiatives, while state capitalism stress the central role of government in political economy.
2 Note the long simmering anti-globalization movement dating back to the 1990s (see Juris 2008) and the current contentious dissatisfaction with markets in light of worsening inequality at a time of vibrant economic growth at both the national and global levels (see Occupy Wall Street n.d.).
3 See the efforts of the members of the International Council on Mining and Metals (n.d.) in promoting sustainable industry practices.
4 These are examples where it is probably better to leave markets alone, where extra-market institutions or interventions become a liability by allowing economic activity to be a weapon or tool in the pursuit of national interests.
5 This entails property ownership over the goods of the earth that one has made productive with one's initiatives and effort (Locke 1988 [1689], sec. 2, para. 27).
6 In claiming rightful property ownership, we are bound by the obligation to leave "enough and as good" in the commons so that others, too, might have a chance to secure such property rights for themselves (Locke 1988 [1689], sec. 2, para. 26, 32).

References

Aquinas, Thomas. 1911–1925. *Summa Theologica*. Translated by Fathers of the English Dominican Province. New York: Benziger Brothers.

Barrera, Albino. 2001. *Modern Catholic Social Documents and Political Economy*. Washington, DC: Georgetown University Press, 2001.

Bator, Francis M. 1958. "The Anatomy of Market Failure." *The Quarterly Journal of Economics* 72: 351–379. 10.2307/1882231.

Burianski, Markus, Damien Nyer, and Federico Parise Kuhnle. 2019. "Resolving Disputes in Africa's Mining Sector." *Africa Focus*. https://www.whitecase.com/publications/insight/resolving-disputes-africas-mining-sector.

Campbell, Greg. 2012. *Blood Diamonds: Tracing the Deadly Path of the World's Most Precious Stones*. New York: Basic Books.

Francis. 2015. *Laudato Si'*. http://www.vatican.va/content/francesco/en/encyclicals/documents/papa-francesco_20150524_enciclica-laudato-si.html.

Frankel, Jeffrey. 2010. "The Natural Resource Curse: A Survey." NBER Working Paper No. 15836. 10.3386/w15836.

Gasper, Des. 2012. "Development Ethics: Why? What? How? A Formulation of the Field." *Journal of Global Ethics* 8, no. 1 (2012): 117–135. 10.1080/17449626.2012.672450.

Goulet, Denis. 2006. *Development Ethics at Work: Explorations 1960–2002*. New York: Routledge. 10.4324/9780203086643.

Humphreys, Macartan, Jeffrey Sachs, and Joseph Stiglitz, eds. 2007. *Escaping the Resource Curse*. New York: Columbia University Press.

Herrendorf, Berthold, Richard Rogerson, and Ákos Valentinyi. 2013. "Growth and Structural Transformation." NBER Working Paper No. 18996. 10.3386/w18996.

Hollick, Ann. 1977. "The Origins of 200-Mile Offshore Zones." *American Journal of International Law* 71, no. 3: 494–500. 10.1017/S0002930000106499.

International Council on Mining and Metals. n.d. "Environment." Accessed December17, 2020. https://www.icmm.com/en-gb/environment.

John Paul II. 1981. *Laborem Exercens*. https://www.vatican.va/content/john-paul-ii/en/encyclicals/documents/hf_jp-ii_enc_14091981_laborem-exercens.html.

Juris, Jeffrey. 2008. *Networking Futures: The Movements Against Corporate Globalization*. Durham, NC: Duke University Press. 10.2307/j.ctv11sn2g6.

Kimani, Mary. 2009. "Mining to Profit Africa's People." *Africa Renewal*. https://www.un.org/africarenewal/magazine/april-2009/mining-profit-africa%E2%80%99s-people.

Leo XIII. 1891. *Rerum Novarum*. https://www.vatican.va/content/leo-xiii/en/encyclicals/documents/hf_l-xiii_enc_15051891_rerum-novarum.html.

Locke, John. 1988 (1689). *Two Treatises of Government*. Edited byPeter Laslett. Cambridge, UK: Cambridge University Press.

Mann, Howard. 2015. *IISD Handbook on Mining Contract Negotiations for Developing Countries. Volume 1: Preparing for Success*. Winnipeg, Canada: International Institute for Sustainable Development. https://www.iisd.org/publications/iisd-handbook-mining-contract-negotiations-developing-countries-volume-1-preparing.

Martinez-Alier, Joan. 2001. "Mining Conflicts, Environmental Justice, and Valuation." *Journal of Hazardous Materials* 86: 153–170. 10.1016/s0304-3894(01)00252-7.

McBeth, John. 2018. "Indonesia Pollution Charge Piles Pressure on Freeport." *Asia Times*, May 9, 2018. http://www.atimes.com/article/indonesia-pollution-charge-piles-pressure-on-freeport.

Mills, Catherine. 2010. *Regulating Health and Safety in the British Mining Industries, 1800–1914*. London: Routledge. 10.4324/9781315244884.

Nest, Michael. 2011. *Coltan*. Cambridge, UK: Polity.

Nordhaus, William D. and James Tobin. 1973. "Is Growth Obsolete?" In *The Measurement of Economic and Social Performance*, edited byMartin Moss, 509–564. New York: National Bureau of Economic Research. https://www.nber.org/books-and-chapters/measurement-economic-and-social-performance.

Norges Bank. n.d. "About the Fund." Accessed January18, 2019. https://www.nbim.no/en/the-fund/about-the-fund.

Occupy Wall Street. n.d. "About." Accessed January18, 2019. http://occupywallst.org/about/.

Okonta, Ike, and Oronto Douglas. 2001. *Where Vultures Feast: Shell, Human Rights, and Oil in the Niger Delta*. San Francisco: Sierra Club Books.

Paul VI. 1967. *Populorum Progressio*. http://www.vatican.va/content/paul-vi/en/encyclicals/documents/hf_p-vi_enc_26031967_populorum.html.

Phelps, Nicholas, Miguel Atienza, and Martin Arias. 2015. "Encore for the Enclave: The Changing Nature of the Industry Enclave with Illustrations from the Mining Industry in Chile." *Economic Geography* 91, no. 2: 119–146. 10.1111/ecge.12086.

Pigou, Arthur Cecil. 2013 (1920). *The Economics of Welfare*. London: Palgrave Macmillan.

Piketty, Thomas. 2017. *Capital in the Twenty-First Century*. Translated byArthur Goldhammer. Cambridge, MA: Belknap Press of Harvard University Press. 10.4159/9780674982918.

Pontifical Council for Justice and Peace. 2004. *Compendium of the Social Doctrine of the Church*. http://www.vatican.va/roman_curia/pontifical_councils/justpeace/documents/rc_pc_justpeace_doc_20060526_compendio-dott-soc_en.html.

Slav, Irina. 2018. "Nigeria Rated Poorest Country Despite Oil Earnings." *OilPrice.com*, July 4, 2018. https://oilprice.com/Latest-Energy-News/World-News/Nigeria-Rated-Poorest-Country-Despite-Oil-Earnings.html.

United Nations Convention on the Law of the Sea. 1982. Montego Bay, December 10, 1982. *United Nations Treaty Series*, vol. 1833, no. 31363. https://treaties.un.org/Pages/ViewDetailsIII.aspx?src=TREATY&mtdsg_no=XXI-6&chapter=21&Temp=mtdsg3&clang=_en.

13 Good governance for mining and the promotion of peace in Africa

Elias O. Opongo, SJ

The increasing frequency and intensity of conflicts in Africa in the last two decades are largely attributed to poor governance amid the rising discovery and exploitation of mining resources (Afful-Koomson 2015, 1). Such conflicts are driven by issues such as access to minerals, revenue distribution, environmental impacts, and corruption, among others. According to the World Bank (2021), extractive industries play a dominant role in the social, political, and economic life of 81 countries with a total population of 3.5 billion people. If the industries are well managed, the populations in these countries should all live above the poverty line, but that is not the case. To a great extent, revenues generated from mining have not been used to support sustainable national development or meet the needs of populations in an inclusive way, especially in the Global South. Consequently, many resource-rich countries in Africa have high poverty levels and unemployment rates, and generally poor economic conditions despite large revenues from the exploitation of minerals.

This mismanagement has impacted conflict in Africa. As Philippe Le Billon (2001, 561) observes, extractive resources "shape strategies of power based on the commercialisation of armed conflict and the territorialisation of sovereignty around valuable resource areas and trading networks." The poor and marginalized, especially women and children, are often the most affected. In 2019, Pope Francis raised concern over systematic injustices that people continue to suffer due to mismanagement in the mining industry, such as land loss, high poverty levels in mining communities, threats and violence, high rates of corruption, and forced prostitution and rape (Gomes 2019). The pope further asserted that, "We need to ensure that mining activities lead to the integral human development of each and every person and of the entire community."

Given these problems, there is an urgent need to promote good governance of mining on the African continent. Governance refers to political and institutional processes that "shape how scientific and technological processes are directed, how environmental and health issues are defined and addressed, and how social consequences are distributed" (Leach et al. 2007). It is crucial to include citizens in the processes of governance in the

DOI: 10.4324/9781003094272-13

mining industry because they tend to be the ones most affected, negatively or positively, as well as directly and indirectly.

This chapter explores the issue of good governance in mining and the promotion of peace, with examples largely from the Great Lakes and Eastern Africa region, but also other parts of the continent. It examines how the governance of mining can be improved through a Catholic peacebuilding approach, founded on Catholic social teaching. The aim is to ensure that natural resources are protected as necessary for the common good; exploited within a framework of care for the environment, protection of human dignity, and sensitivity to the impact on conflict dynamics; and used for revenues that are distributed equitably to the majority of the population.

Contextualizing mining industry governance in Africa

Africa has a wide range of mineral deposits used in producing jewelry, automobiles, electronics, renewable batteries, and other goods. As a result, mining generates significant income, but Africa's enormous wealth of natural resources has yet to translate into significant improvement in people's living standards. The mining sector is marred by high levels of corruption and criminality, perpetrated by armed groups who use proceeds from the sale of minerals like gold, tantalum, and tin to finance violent activities (Lebert 2016). Resource-based conflicts have occurred in Kenya, the Democratic Republic of the Congo (DRC), Burundi, the Central African Republic, Liberia, South Sudan, and elsewhere. Also, companies and individuals that engage in mining activities frequently do not take the necessary steps to ensure transparency in the supply chain. Furthermore, many people across the region engage in artisanal mining to earn a living despite the dangers of exposure to human rights abuses, corruption, or violence (White and Beck 2017). Artisanal and other types of mining also tend to destroy the environment and expose communities to further ecological vulnerabilities, such as polluted water sources, open pits that could be harmful to the community, and soil erosion.

The social-economic contradictions in many parts of Africa translate into a paradox of suffering in the midst of plenty. Rich mineral resources largely benefit foreign multinational companies (MNCs), elites within national governments, and insurgent groups, while leaving the rest of the population poor and desperate. This tends to contradict the economic theory that holds that foreign direct investments "create growth multiplier effects through vertical and horizontal spillover effects: including the transfer of technology and know-how to domestic firms, the formation of human capital, etc" (Colen, Maertens, and Swinnen 2013, 23). Poor governance, while not the only culprit, is a central reason why these economic benefits do not come to fruition. It allows the MNCs that play a major role in mineral supply chains to engage in exploitative activities that promote crime, corruption, and inequitable sharing of mineral wealth, which fuels violent conflicts.

Kenya has experienced numerous resource-induced conflicts, largely due to competition for land and water for both human livelihood and livestock grazing, but also for control over mines. For example, the residents of Kwale County in the coastal region of Kenya have been opposed to titanium mining while citing concerns over displacement of the local populations, and pollution of water resources that the government and the Canadian mining company failed to address (Greyl 2019). In addition, some of the communities that were forced to cede land for mining have not been compensated. The government's discriminatory approach—compensating some but not all communities—has led to regional conflict. There have also been cases where competition for land used for small-scale mining of sand and special stones has led to community conflicts.

The DRC has immense natural resource wealth, but with poor governance structures for overseeing them, the country has become "a textbook forum for natural resource-induced conflicts at both local and national levels" (Burnley 2011, 7). The country is among those that have experienced the bloodiest conflicts since the Second World War, conflicts that often have been over control of mineral resources, as Rigobert Minani describes in this book (see also Snow 2013). As elsewhere in Africa, the DRC's enormous wealth of natural resources has yet to translate into significant improvement in people's living standards. For instance, the poverty level is one of the highest on the continent. With its 2019 Human Development Index rating of 0.48, which is below the average of 0.513 for countries within the low development group, the DRC was ranked 175 out of 189 countries (United Nations Development Program 2020, 2). While the DRC has fifty percent of the forests and water resources in Africa, the country's environment is negatively affected by "deforestation, species depletion, heavy metal pollution and land degradation from mining, as well as an acute drinking water crisis which has left an estimated 51 million Congolese without access to potable water" (United Nations Environment Programme 2017, n.p.). Ninety percent of mining is artisanal (Samndong and Nhantumbo 2015, 6). Landscapes in the mining regions have generally been destroyed, while many people in these regions live in deplorable conditions due to extreme poverty and poor delivery of crucial government services (EU-UN Partnership on Land, Natural Resources and Conflict Prevention 2015). As Minani describes, a major push from Catholic groups and others led to a revision of the country's mining contracts, but a lack of transparency and oversight when the contracts were finally renegotiated left the country in the dark about how governance would actually be improved to address these challenges.

Other countries have experienced similar problems tied to mining and other extractives. Sierra Leone is a well-known example, where diamond mining was used to fund militia groups. These groups fought over control of this resource, while using revenues to purchase arms. South Sudan has large oil deposits but is one of the world's poorest countries. The country has been mired in conflict since 2013, leading to civilian displacement and death. The conflict has disrupted oil production, the main source of government revenue. There has been destruction of forests for firewood and charcoal, and artisanal miners have degraded the environment searching for diamonds and gold. The Malawi Ministry of Energy, Natural Resources and Environment (2010, iii) has cited challenges in the country's extractive industry similar to those in DRC: "deforestation; decreasing soil fertility and increasing erosion; water depletion; loss of biodiversity; and increasing pollution." Likewise, Tanzania has been battling an increasing problem of coastal erosion attributed to sand mining (Gavriletea 2017).

These cases demonstrate how resource-based conflicts, to a great extent, are attributable to poor management in the mining industry. Luca Ventura (2020, n.p.) argues that all "extremely fragile and underdeveloped economies have either recently been through a civil war or are suffering from ongoing sectarian or ethnic conflicts." Mineral-rich countries in Africa fit this pattern. Hence, good and efficient extractive management policies and structures are critical for sustainable economic development that can overcome those adverse conditions and lead to peace and stability. As noted above, in Africa, conflicts have largely been facilitated by unfair distribution of natural resources and marginalization of large parts of the population in the course of exploitation of these resources. According to Clementine Burnley (2011), countries that have quality institutions dedicated to the management of natural resources have a low likelihood of plunging into violent conflicts. Institutional structures and norms like constitutional order, equal justice under law, democratic principles, independent judicial and legislative systems, and executive accountability—when enshrined in law and practice—diminish chances of social conflicts and injustices related to mining. For example, in South Sudan large quantities of gold are smuggled out of the country on a regular basis, largely to benefit the political elite, business people, and foreign investors (Mach 2017). The proceeds from the smuggled minerals tend to encourage the formation of militant groups that can destabilize a nation. This explains why the mineral-rich area of eastern DRC has the largest number of militia groups and the worst forms of human rights abuses in the region. Poor governance allows patterns like this to exist, and improved governance is the key to curtailing them.

Distinctive aspects of a Catholic peacebuilding approach

A Catholic peacebuilding approach comprises a "broad spectrum of activities, a wide variety of actors, and diverse elements of Catholic beliefs and teachings that must come together for the Catholic community to fulfill its mission as a Peacebuilding Church" (Powers 2017, 1). A Catholic peacebuilding approach is anchored in the Catholic social teaching principles of common good and human dignity. These two principles ground the Catholic commitment to safeguard the environment, secure livelihoods for current and future generations, and responsibly manage the benefits accrued from the earth's resources so that the entire population of a country is not only a stakeholder, but also a beneficiary. Such an approach ensures peace and stability in the country.

Incarnations of Catholic peacebuilding vary across Africa. For instance, in 2018 in the DRC, the church oversaw mediation processes and the implementation of political pacts, and played a vital role in voter education and advocating for free and fair elections. In South Sudan, the South Sudan Council of Churches and other church actors have been instrumental in peacebuilding efforts: pressing for peace negotiations, sponsoring civil society peace processes, and focusing on advocacy and dialogue through impartial forums that have fostered reconciliation. This culminated in Pope Francis' invitation to the key protagonists in the conflict to come to the Vatican for a reconciliation retreat in April 2019. The pope kissed the feet of the political leaders and begged them to end the conflict and embrace peace. Subsequently, in addition to other diverse and complementary efforts for peace, the Sant'Egidio Community, a lay Catholic group based in Rome, has facilitated peace negotiations involving President Salva Kiir, his opponent and former deputy Riek Machar, and other actors. These negotiations have led to agreements that have allowed for the formation of a government of national unity, bringing an end to intermittent conflicts that have plagued the youngest nation in the world since 2013. The pope's intervention showed the church's commitment to peace and conflict resolution. Additionally, Pope Francis's visit to the Central African Republic in 2017, at the peak of civil violence, to appeal for peaceful resolution to the conflict was a heroic gesture that demonstrated the church's solidarity with the suffering population. In Nigeria, Burkina Faso, Eritrea, Zimbabwe, Kenya, and South Africa, among many other countries, the church has strongly spoken against exploitation of the poor, corruption, poor governance, and extreme forms of violence by militia and terrorist groups. The church, in its prophetic mission to defend human dignity and safeguard the common good, has largely been influenced by the Catholic peacebuilding approach.

A Catholic peacebuilding approach is distinctive in two main ways that relate directly to issues of mining and governance. First is the emphasis that the church puts on "the biblical conception of peace, the centrality of peace in the Church's sacraments and mission, and a distinctively Christian theological and ethical approach to war and peace" (Powers 2017, 4). The theological approach to peace begins in Scripture, and Cardinal Peter Turkson provides a reflection in this book on how the Bible colors a Catholic understanding of mining. He describes how metals and minerals mined from the earth are, like all creation, intended by God to contribute to the building up of the heavenly kingdom. The goods of the earth are meant to be for more than just profit or utility, but a peaceful, rightly ordered human society builds upon economic development. Hence, good governance of mining and promotion of peace, rooted in equitable sharing of common resources and a deliberate option for the poor, are grounded in a normative theological foundation. Mining that creates marginalization, environmental destruction, and threats to livelihoods, and that in turn feed violent conflict, ought to be challenged and critiqued in light of that normative ethic. For Catholic actors and organizations working in the mining sector, this theological foundation should help inform and motivate them. And for Catholic ecclesial leaders it should animate their pastoral leadership and drive them to engage the faithful, as well as their broader national and regional communities, with an understanding that mining issues are crucial to the just peace that marks the kingdom of God.

A second distinctive aspect of a Catholic peacebuilding approach is the church's use of the tradition of natural law, which is based on the presumption that "the demands of justice and peace are knowable and binding on all persons, regardless of their faith commitments or lack thereof" (Powers 2017, 4). This approach embraces all humanity, regardless of faith or socio-cultural affiliation. At the center of the natural law interpretation is the safeguarding of human dignity and the common good. Hence, the inhuman circumstances under which some mining activities operate are against the natural law of respect of human dignity. Peacebuilding in the mining industry through advocacy for fair distribution of the benefits of mining, sustainable environmental practices, safe working conditions, and concern for future generations is an outgrowth of natural law. Such an approach can help Catholic peacebuilders reach outside their own tradition and engage other faith traditions, secular groups, and governments and corporations on common issues of concern related to ethics and good governance in the mining industry. Catholic peacebuilding "calls on Christians to cooperate with all people in securing a peace based on justice and love" (Curran 1984, 61).

Catholic social teaching and governance of extractive industries

Catholic social teaching (CST) can provide guidance for governance of mining industries in Africa. When applied to mining, these principles include the following assumptions: natural resources are common goods; political authorities have a right and duty to manage these resources in transparent and equitable ways that respect human rights and do, in fact, contribute to the common good; the environment must be protected in the course of exploitation of natural resources; natural resources should be managed in ways that mitigate conflict and promote peace; and the principles of justice and solidarity require that the imbalance of power between host African countries and MNCs should be rectified.

CST emphasizes management of natural resources as common goods that ought to be enjoyed by all. This emphasis has two main dimensions: preservation of common resources for current and future generations, as well as fair distribution of benefits. The two imperatives are co-existent and complementary. On preservation of common resources, Pope Francis (2015, §93) states that "the earth is essentially a shared inheritance, whose fruits are meant to benefit everyone. For believers, this becomes a question of fidelity to the Creator, since God created the world for everyone." Inclusion of future generations in the planning and exploitation of natural resources is a crucial part of this obligation to preserve common resources. Since natural resources are a common inheritance, the costs and benefits of their extraction ought to be shared fairly among the population. Their exploitation should not be for the financial gain of the few at the expense of the many. Communities living around mines ought to be protected from possible negative effects on their well-being.

Transparency in the management of the financial wealth and profits accruing from mining should be maintained as a vital principle in extractive governance. The common good cannot be protected or strengthened if corruption cannot be identified and if bad financial deals prevent the people from receiving just benefits from the country's mineral resources. Transparency also makes space for communities near mining sites to be more engaged, making for more possibility of consensus building among all stakeholders. Such consensus is important for promotion of human dignity, peace, and stability. Pope Francis (2015, §183) affirms that "the local population should have a special place at the table; they are concerned about their own future and that of their children, and can consider goals transcending immediate economic interest."

While exploitation of natural resources for the benefit of all citizens in a country is acceptable, care should be taken to ensure that the environment is preserved and quality of human life safeguarded. Good mining management practices require minimizing negative effects on the environment, whether on or off-site, and undertaking mitigation and rehabilitation procedures for ecological damage. Of major concern in many African countries is the lack of institutionalized policies that are operationalizable within the framework of mining, especially for ensuring benefit to the majority of the population and protecting the environment.

In most mining settings, as pressure mounts on the environment and the chances of conflicts increase, the poor are especially vulnerable. In this book, Holden and Montevecchio examine the issue of climate justice—that is, the way demand for minerals in the Global North causes environmental harm that disproportionately affects poor and marginalized peoples in the Global South. It also explains how mining is one of the leading correlates with instances of violent conflict related to such increased environmental vulnerability. This reality is what led Pope Francis (2015, §48) to observe that "both everyday experience and scientific research show that the gravest effects of all attacks on the environment are suffered by the poorest." Bad management of the environmental consequences of mining increases levels of poverty, worsening living standards, and rendering communities vulnerable to conflicts.

If mining is properly regulated and managed, it can contribute to alleviating poverty and underdevelopment while protecting the environment. But, as Pope Benedict XVI (2011, §79) noted, unjust and exploitative economic activities by the elites should not be disguised as actions that seek to eradicate poverty. Pope Francis (2015, §51) diagnosed how multinational companies exploit the economic disadvantage of countries in the Global South with damaging practices that they would not carry out in their home countries. Such patterns both feed off and reinforce global power inequalities. They inhibit sustainable development, damage the environment, and threaten peace and stability. According to Francis, the spirit of solidarity, which serves as a critique of this kind of exploitation, must be institutionalized in strong and effective governance structures, laws, and policies (§142).

Harnessing the peacebuilding capacity of the Catholic Church in mining industries

The church can increase its potential for peacebuilding as a strategy for addressing issues around governance of mining if it strengthens three ecclesiastical characteristics: being technically informed; being communally present; and being engaged and committed to social change. This will require ecclesiastical and pastoral strategies that strengthen the church's capacity to effect social-political change for the benefit of the people in an area that requires significant specialized knowledge, prolonged dedication, and a varied coalition.

A "technically-informed church" refers to the need for educating church leaders and laity about the technical aspects of mining industries. The complexities of exploration, environmental assessments, means of exploitation, community development agreements, systems for distribution of benefits from extractives, environmental impact mitigation, and exit strategies are a clear indication that any involvement in this industry presupposes a high level of knowledge on key technical issues. Raymond Offenheiser, in his chapter, argues that in order to be taken seriously by industry leaders and be effective change agents, Catholic peacebuilders need to ally with people that have technical expertise and work to expand their own technical knowledge in these areas. Thus the church and lay organizations need funding, dedicated staff, and leaders who are committed to working on the issue for a long period of time. Technical expertise will also require liaising with local and international experts in the field of mining within Catholic and non-Catholic networks, education institutions, and civil society networks. Organizations like The People's Dialogue[1] and the Thematic Social Forum on Mining and the Extractivist Economy,[2] Haki Madini ('justice for minerals' in Kiswahili) in Tanzania,[3] Women in Mining National Network Structure (RENAFEM in French) in the DRC, Extractive Sector Forum (ESF) in Kenya, and others are crucial players in mining industry advocacy and have developed robust networks of technical knowledge.

Technical knowledge is also important in contract negotiations between MNCs and host African countries. The process of drafting and signing these contracts is often questionable. Most African governments do not have the technical know-how on fair negotiation, monitoring of the exploration and exploitation of resources, and the technology and practices required to ensure that the populations remain safe from any possible negative effects in the short, medium, and long term. Pope Francis (2015, §175) emphasizes that "The twenty-first century, while maintaining systems of governance inherited from the past, is witnessing a weakening of the power of nation-states, chiefly because the economic and financial sectors, being transnational, tends to prevail over the political." He adds that we must create international mechanisms for management of natural resources that cushion poorer countries from negative impacts of extractive industries while promoting transparency and accountability. But such improvements require increased knowledge and capacity, and the church needs to lead the way by supporting the education of its communities and using that knowledge for effective policy advocacy for governance that can better withstand the huge power imbalances that exist between MNCs and host countries. The Catholic organizations that carried out the study of DRC mining contracts described by Minani serve as a good model for how other Catholic actors can take on this very technical, but very important, aspect of mining governance.

The second ecclesiological characteristic is that the church ought to be present to its people. In most cases, mining activities are carried out in isolated and marginalized locations, or in situations of protracted conflict. This can exhaust the church leaders to the extent of abandoning the people. Maintaining stamina to stay in solidarity with the people and walk the journey of justice with them is fundamental. This challenge makes the church's ability to connect advocacy around mining to core elements of its spiritual and pastoral practice vitally important. In this book, the chapter by Polanía-Reyes and Henao on Colombia offers a strong positive example. The Colombian Church has identified mining and ecology as key dimensions of the implementation of its national peace process, and is encouraging a stronger presence of ecological theology in its pastoral mission so as to make ecclesial leaders and systems more firmly invested in how mining and ecological damage present challenges to peace and development.

The third characteristic is direct engagement on social justice issues. This presupposes the two previous characteristics: that there is a high level of knowledge on technical issues in the mining sector, and that the church is effectively present in mining communities so that it understands their challenges and needs. This third characteristic is focused on channeling the other two into specific actions for improving governance of natural resources. It is important that church actors use technical knowledge and understanding gleaned from accompanying local communities to conduct contextual analyses that identify major issues of concern and strategize on how to address them. Not every instance of mining presents the same challenges. For example, different companies conduct their business in different ways, different minerals demand different techniques and have different impacts, different countries have different cultures and histories that prioritize different values, and different countries and regions have to deal with their own unique conflict factors. There are no one-size-fits-all solutions, and so the church needs enough savvy and expertise to develop specific policy priorities and advocacy agendas. Therefore, church groups need to establish advocacy networks with diverse organizations—NGOs, civil society groups, different faiths, and inter-government institutions. These networks can bring positive change by promoting the common good, protecting human dignity and safeguarding human livelihoods and the environment. These social justice activities bring out the servant leadership aspects of the church.

While justice priorities will vary as described, the management of natural resources in situations of conflict is an important cross-cutting factor given the fact that many countries in Africa have experienced resource-based conflicts. Hence, there should be clear links between conflict resolution strategies and advocacy on justice and environmental issues related to the exploitation of natural resources. Pope Francis (2015, §92) reiterates that "Peace, justice and the preservation of creation are three absolutely interconnected themes, which cannot be separated and treated individually without once again falling into reductionism." Different mechanisms of managing and preventing both existing and potential conflicts in extractive industries are crucial for sustainability of the sector. These include integration of conflict resolution measures into development agreements between communities, MNCs, and governments, as well as into broader governance structures of mining. Equally important is that environmental impact assessments ought to anticipate the potential impact of natural resource exploitation on conflicts. Conflict prevention also requires fair distribution of the benefits of mining, especially to the communities most affected.

Conclusion

Africa is home to close to thirty percent of the world's minerals and has great potential to achieve peace and prosperity. However, poor governance of extractive resources has rendered the continent vulnerable to external exploitation by MNCs, as well as the corruption and greed of political and social elites who have continued to deny the majority of the population the benefits of mining. In addition, mining has contributed to environmental degradation and inter-communal conflicts. The church has a major role in advocating for institutional reforms within government and the mining industry in order to achieve good governance and accountability, and it is well-equipped to do so. It must continue to insist that natural resources are a heritage that ought to be safeguarded as common goods to be enjoyed by all, while taking into account the needs of future generations and peace in affected communities and nations.

Notes

1 See https://www.peoplesdialogue.org/.
2 See https://www.thematicsocialforum.org/.
3 See http://hakimadini.org/.

References

Afful-Koomson, Timothy. 2015. "Collaborative Governance and its Relevance for Extractive Industries in Africa." In *Collaborative Governance in Extractive Industries in Africa*, edited by Timothy Afful-Koomson and Kwabena Owusu Asubonteng, 1–26. Accra: United Nations University Institute for Natural Resources. 10.2307/j.ctvh9vtth.

Benedict XVI. 2011. *Africae Munus*. http://www.vatican.va/content/benedict-xvi/en/apost_exhortations/documents/hf_ben-xvi_exh_20111119_africae-munus.html.

Burnley, Clementine. 2011. "Natural Resources Conflict in the Democratic Republic of the Congo: A Question of Governance?" *Sustainable Development Law & Policy* 12, no. 1: 7–11, 52–53. https://digitalcommons.wcl.american.edu/sdlp/vol12/iss1/5/.

Colen, Liesbeth, Miet Maertens, and Johan Swinnen. 2013. "Foreign Direct Investment as an Engine for Economic Growth and Human Development: A Review of the Arguments and Empirical Evidence." In *Foreign Direct Investment and Human Development: The Law and Economics of International Investment Agreements*, edited by Olivier De Schutter, Johan Swinnen, and Jan Wouters, 70–115. London: Routledge. 10.4324/9780203076880.

Curran, Charles E. 1984. "Roman Catholic Teaching on Peace and War Within a Broader Theological Context." *The Journal of Religious Ethics* 12, no. 1: 61–81. https://www.jstor.org/stable/40014969.

EU-UN Partnership on Land, Natural Resources and Conflict Prevention. 2015. "The Great Lakes Region: A Pilot Project." https://www.un.org/en/land-natural-resources-conflict/pdfs/eu-un-great-lakes-project.pdf.

Francis. 2015. *Laudato Si'*. http://www.vatican.va/content/francesco/en/encyclicals/documents/papa-francesco_20150524_enciclica-laudato-si.html.

Gavriletea, Marius Dan. 2017. "Environmental Impacts of Sand Exploitation: Analysis of Sand Market." *Sustainability* 9, no. 7: 1118. 10.3390/su9071118.

Gomes, Robin. 2019. "Pope: Mining Activities Should Ensure Integral Sustainable Human Development." *Vatican News*, May 3, 2019. https://www.vaticannews.va/en/pope/news/2019-05/pope-francis-mining-indigenous-rights-development-common-home.html.

Greyl, Lucie. 2019. "Titanium Mining in the Kwale District, Kenya." Environmental Justice Atlas. Updated August 18, 2019. https://ejatlas.org/conflict/titanium-mining-in-the-kwale-district-kenya.

Le Billon, Philippe. 2001. "The Political Ecology of War: Natural Resources and Armed Conflicts." *Political Geography* 20, no. 5: 561–584. 10.1016/S0962-6298(01)00015-4.

Leach, Melissa, Gerald Bloom, Adrian Ely, Paul Nightingale, Ian Scoones, Esha Shah, and Adrian Smith. 2007. "Understanding Governance: Pathways to Sustainability." STEPS Working Paper 2. STEPS Centre, May 20, 2007. https://steps-centre.org/publication/understanding-governance-pathways-to-sustainability/.

Lebert, Joanne. 2016. "Conflict-Prone Minerals in the Great Lakes Region of Africa." *Partnership Africa Canada*, January 1, 2016. https://www.africaportal.org/publications/conflict-prone-minerals-in-the-great-lakes-region-of-africa-2/.

Mach, Parach. 2017. "War-torn South Sudan Gold Goes to Dubai: Official." *Anadolu Agency*, March 3, 2017. https://www.aa.com.tr/en/africa/war-torn-south-sudan-gold-goes-to-dubai-official/763633.

Malawi Ministry of Energy, Natural Resources and Environment. 2010. *Malawi State of Environment and Outlook Report*. Lilongwe, Malawi: Malawi Environmental Affairs Department. https://www.ead.gov.mw/storage/app/media/Resources/Reports/Malawi%20State%20of%20the%20Environemnt%20and%20Outlook%20Report_2010.pdf.

Powers, Gerard. 2017. "Toward an Integral Catholic Peacebuilding." *The Journal of Social Encounters* 1, no. 1: 1–13. http://digitalcommons.csbsju.edu/social_encounters/vol1/iss1/2/.

Samndong, Raymond Achu, and Isilda Nhantumbo. 2015. *Natural Resources Governance in the Democratic Republic of Congo: Breaking Sector Walls for Sustainable Land Use Investments*. IIED Country Report February 2015. London: IIED. https://pubs.iied.org/sites/default/files/pdfs/migrate/13578IIED.pdf.

Snow, Dan. 2013. "DR Congo: Cursed by Its Natural Wealth." *BBC News*, October 9, 2013. https://www.bbc.com/news/magazine24396390.

United Nations Development Program. 2020. "Human Development Report 2020 Briefing Note for Congo (Democratic Republic of the)." Accessed May 13, 2020. http://hdr.undp.org/sites/all/themes/hdr_theme/country-notes/COD.pdf.

United Nations Environment Programme. 2017. "Study Confirms DR Congo's Potential as Environmental Powerhouse but Warns of Critical Threats." August 7, 2017. https://www.unenvironment.org/fr/node/848.

Ventura, Luca. 2020. "Poorest Countries in the World 2020." *Global Finance*, July 22, 2020. https://www.gfmag.com/global-data/economic-data/the-poorest-countries-in-the-world.

White, Natasha, and Lucy Beck. 2017. "Only Half of African Great Lakes Mineral Exporters Reported on Responsible Sourcing Efforts: New Global Witness Report." *Global Witness*, August 30, 2017. https://www.globalwitness.org/en/press-releases/only-half-african-great-lakes-mineral-exporters-reported-responsible-sourcing-efforts-new-global-witness-report/.

World Bank. 2021. "Extractive Industries: Overview." Updated April 26, 2021. www.worldbank.org/en/topic/extractiveindustries/overview.

14 Mining and the call for solidarity: the networks we have and the synodal network the Church is called to be

Vincent J. Miller

The details are painfully familiar: extractive zones scarred by ecological destruction, local populations suffering from poisoned water and air as well as social and cultural disruption, while local and national governments are corrupted by the cash the extracted commodities generate. Thousands of miles away, consumers enjoy the benefits of products made by these resources, rarely giving thought to the costs borne by others because they are largely unaware of them.

That lack of connection is not simply a moral blind spot, but a result of the powerful global market network that constructs and constrains relationships in which we are all viscerally enmeshed. Analyzing this as a network problem as opposed to a failure of moral will provides a more complex understanding of the problem and better guidance for contesting it. The church can also be considered a network with its own vision for how relationships should be structured. The relationships fostered by the church are currently weaker than global market relations. From a theological perspective, we are united to one another in Christ; but practically, Christians relate to fellow church members in distant lands primarily through the webs of the market. The mystical Body of Christ is too often bound together by the sinews of global capitalism. The church not only can, but, to remain true to its mission, must provide an alternate to the constrained relationships constructed by the global economy.

This essay will seek to chart ways in which the church can engage the mining practices that undergird industrial society. First, it will consider the power and history of the network of the global market in raw materials; then, it will review the standard means for influencing these markets and the mining projects that feed them; and it will conclude by proposing additional means for challenging the injustice of these projects that emerge when the church's capacity as a global network is taken seriously. As the Synod on the Pan-Amazonian Region demonstrated, the church can do more than exhort its members to exercise solidarity. It can function as a global network to facilitate deeper relationships of solidarity that challenge the shallow and often exploitative relationships fostered by contemporary capitalism.

DOI: 10.4324/9781003094272-14

The network we have

Contemporary globalized capitalism comprises a network that reaches from the most developed cities to the least developed locations on earth. This is the most powerful and robust network that has ever existed, able to move valued resources from mines far from transportation hubs, energy sources, or skilled labor into global commodity chains that feed production in industrial centers. It is astoundingly resilient when compared to other essential networks such as education, health, or rule of law. Regardless of battles over territory, legality or illegality, sustainability or unsustainability, peace or war, strong, weak or failed state, minerals flow to manufacturers and products get made.

This economic network is heir to centuries of colonial extraction. Theologian Daniel Castillo (2019) offers a concise synthesis of critiques of the colonial legacies hidden within contemporary extractive economies. He describes a "500 year project" in which the systems of colonial plunder became the basis for the capital accumulation that funded industrial society and Western modernity (142). Uruguyan Eduardo Galeano (1997; cited in Castillo 2019, 150), whom Pope Francis cites in *Querida Amazonia*, provides a contemporary witness to this expropriation: "Everything, from the discovery until our times, has always been transmuted into European—or later United States—capital, and as such has accumulated in distant centers of power." Castillo 2019 The Catholic Church confessed its complicity in this process during the Synod of Bishops Special Assembly for the Pan-Amazonian Region (2019, §15) in "The Amazon: New Paths for the Church and for an Integral Ecology" : "The proclamation of Christ often took place in collusion with the powers that exploited the resources and oppressed the local populations."

De las Casas' (1986, 2:57; cited and translated in Castillo 2019, 150–51) *Historia de los Indios* provides a colonial-era witness to this extractive regime:

> Mountains are turned over from the bottom to top and top to bottom a thousand times; [the workers] dig, split rocks, move stones, and carry the earth on their backs to wash it in the rivers, and those who wash the gold stay in the water all the time with their waists bent so constantly it breaks their bodies.Castillo 2019

Add coltan, cassiterite, and a handful of other minerals; update the risks to include mercury and cyanide exposure; and de las Casas's description applies to countless extractive sites around the world today.

Castillo's analysis goes beyond tracing how the global economy continues the human injustices of colonialism. He argues that the legacies of colonialism lie within the very logic of extraction. The "technocratic paradigm" (the view that the world is a passive object open to human manipulation and exploitation) that Francis (2015b, §106) decries in *Laudato Si'* has its roots in the colonial plunder which turned societies and cultures, forests and landscapes into sites of extraction and expropriation (Castillo 2019, 27–28, 146–61). The problems we face with contemporary extractive practices thus involve questions of injustice and violence both for those whose lands and livelihoods they impact, and for the ecosystems which they despoil, disrupt, and displace.

But these colonial legacies do not describe all the effects of the global mining market. Contemporary mining is often undertaken voluntarily, by a variety of actors, from artisanal miners and medium-scale lightly mechanized enterprises, to the mega-scale projects of major mining corporations. The balance of positives and negatives in these arrangements vary greatly. Artisanal mining has long been a part of the livelihoods of peoples located near easily accessible mineral sources. Medium-scale enterprises using cheap yet powerful earth-moving, washing, and refining processes can provide substantial wages for impoverished workers, but have very significant environmental impacts and are often tied to illegal actors. Major mining projects are often welcomed by national governments. Some producers have taken great pains to develop non-polluting processes and to negotiate with affected communities. Other large projects exploit weak regulatory regimes to extract maximum profit with little concern for costs borne by local residents and ecosystems.[1] Economist and theologian Daniel Finn (2017) argues that justice-oriented activists should distinguish the operations of good actors from destructive ones and avoid blanket prophetic denunciation of the entire mining industry.[2] There are responsible and destructive actors on all scales.

Both of these perspectives on the nature of mining are important. But here we are concerned with a different problem: how the contemporary structure of the global commodities market provides an outlet for raw materials regardless of the morality of their production. It is on this level that many systemic problems appear (see Arboleda 2020). The existence of legitimate global commodity markets has destabilizing effects because it also provides an outlet for illegal and unregulated extraction, undercutting efforts of good actors. This market is enormously adaptive, able to work through catastrophes and conflict. Villages get burned, populations are terrorized, activists are murdered, miners forced from territory, middlemen killed in turf wars, corporations gain and lose concessions, and still raw material supply flows unabated. As Rigobert Minani documents in his chapter on the Democratic Republic of the Congo, if there is a valuable mineral or resource that can be had, it will find its way into the global commodity market.

It is helpful to consider how this network's material structure provides the basis for certain ways of knowing, acting, and relating that construct and constrain our moral agency in ways we seldom realize. Here we will consider three interrelated aspects: market externalities, commodity distancing, and the price signal as a shallow form of relationship.

One of the fundamental aspects of this global network is the problem of "externalities," costs associated with production that do not appear in the exchange price. These include pollution; destruction of watersheds, agricultural land, and local ecologies; exploitative and destructive labor practices; and physical and cultural violence against local populations. Externalities arise from the immorality of market actors who seek to maximize their profit by shifting costs to others and because any extractive undertaking has many consequences that go unremarked and might only manifest years after a project ends. The latter are inherently difficult to include in the price of exchange. Doing so requires stable and robust state and international regimes with the power to enact and enforce environmental and other regulations requiring producers to minimize pollution and exploitation—what Barrera calls "extra-market interventions" in his chapter in this book. Given the profound adaptability of markets, it is enormously difficult to enact such regulations on a global scale.

Thomas Princen (2002) has coined the term "commodity distancing" to describe how globally dispersed supply chains transform markets. As geographical distance increases, our knowledge about the context and consequences of products we consume decreases. At each point of exchange, only information relevant for the purchase—usually only quality and price—are passed along. This is exacerbated by cultural distance. End consumers have little or no knowledge of the predicaments of farmers, laborers, and communities at the other ends of supply chains, whether this is their political, cultural, economic, and legal situation or the ecological impact of production on the rest of their lives. Princen's comment about consuming Chilean grapes applies to any global commodity: "I have no way of knowing if my consumption is supporting or undermining that farmer, economically or ecologically. With no feedback, or uncertain feedback, I can only assume and, as with most of us, prefer to assume, that my purchases are supportive. I can only assume that all parties have entered the bargain voluntarily and that coercion or extreme dependency have not been the conditions of my enjoyment" (118). Princen's analysis reveals how this global network moves commodities and capital, but not all relevant information about them. This massive filtering of information facilitates the externalization of costs.

Finally, we must consider how economic structures have broader cultural impacts. This moves us out of the realm of economics and into consideration of cultural effects of market structures and how they elicit certain ways of being and relating. Market relations in which the primary form of communication is the price signal are extremely narrow forms of relationship. This is the source of the profound "efficiencies" that market structures enable. Focusing on one aspect of a relationship between parties—the price—enables the massively complex systems of production that characterize modern economies. Each party has a concrete measure by which to evaluate their interaction with others that can be managed through financial accounting.

We need to consider how this powerful economic practice constructs and constrains our relationships. To stay in business, market actors must focus on the price signal. They may prefer one provider's business, labor, and ecological practices, or may prefer to purchase from a local firm, but if the cost is significantly higher, they will not be able to do so, lest their own products' prices become uncompetitive. The price signal disciplines and subordinates other value considerations. Thus, the global economy functions as a massive network of shallow relationships. It facilitates myriad relationships, but constructs them in an extremely narrow manner. Sometimes these are explicitly negotiated relationships between parties, but for the most part, these are implicit but real relationships within supply chains that contribute to the myriad products we consume.

Means of influencing these networks

These interrelated aspects of the global market system are not incidental side effects, but consequences of its central strengths. Thus, our attempts to counter these negative impacts require engaging the dynamics of the global economy itself. There are many methods of influencing these networks to limit their destructiveness and bring external costs into the market price. These include national legislation and international and sector agreements to minimize the actions of immoral actors and support moral ones. Two important policy foci involve transparency and the consent of those affected by extractive projects.

Transparency policies address problems caused by commodity distancing: the tendency of global commodity markets to obscure the origins of industrial materials. These seek to make supply chain accounting public by requiring firms to disclose their negotiations and payments regarding mining projects. One example is the Extractive Industry Transparency Initiative (EITI), a set of voluntary guidelines through which extractive companies and governments disclose all payments for audit and agree to work in partnership with civil society groups (see Lujala 2018; Bamat, Chassy, and Warne 2011, 15). Transparency can also be applied to manufacturers' supply chains. The most well-known manufacturing transparency regimes concern "conflict minerals" (tin, tantalum, tungsten, and gold) derived from mining in the Democratic Republic of the Congo. These include Section 1502 of the 2010 Dodd-Frank Wall Street Reform and Consumer Protection Act, which required publicly traded companies in the US to certify that their supply chains do not source minerals from conflict zones,[3] and the OECD Due Diligence Guidance for Responsible Supply Chains of Minerals from Conflict-Affected and High-Risk Areas, which provides similar regulations for European firms.

The 2008 United Nations Declaration on the Rights of Indigenous Peoples (2007) articulated the notion of "free prior informed consent" (FPIC) in all matters in which the rights, culture, and lands of indigenous people are affected by governments and businesses. Article 32.2 applied this to "any project affecting their lands or territories and other resources, particularly in connection with the development, utilization or exploitation of mineral, water, or other resources." This principle has been generalized to apply to any local community significantly impacted by mining, oil, gas, or any other major project that transforms land or has a significant lasting impact (Bamat, Chassy, and Warne 2011, 10). Communities near mining projects are often rural, poor, and politically and culturally marginalized. FPIC provides a standard to guide political and economic actors to address affected communities' concerns, seeking their consent rather than exploiting their marginality. FPIC has come under criticism as a framework easily manipulated by powerful mining firms against poor and marginalized communities and for being biased toward approval of extractive projects. In response, indigenous, religious, and NGO groups are advocating for "the right to say no"—to allow local communities to simply refuse such operations in their territories.[4]

This is important and often quite effective work. But it is essential to note how fragile these interventions are in contrast to the network itself. This is not simply because of the profound power of wealthy industry actors whose resources dwarf those of NGO and state actors who seek to regulate extractive industries. It has to do with a more fundamental asymmetry in the network itself. To mix a metaphor, market actors know the loopholes they need to operate, whereas those seeking to regulate a market must mind the entire dike of the global supply chain. Again, this problem is a direct consequence of the nature of market exchange. Market actors need only know the immediate local context of exchange: whether something can be sold at profit. While this can involve the strategic advantages of massive transnational corporations, often enough, it is equally a matter of very small-scale informal and black-market actors.

Given the breadth of global demand and the complexities and opacities of the many levels of manufacturing, local actors will almost always find a black-market outlet for destructively mined minerals. Think of the countless uses of gold—from electronics to jewelry. Big, publicly traded, transnational firms will work hard to certify that their supply chains are free of conflict minerals. But there are thousands of sites around the world where small-scale miners are destroying watersheds and ecologies with power-washing, mercury, and cyanide, including sites where illicit armed actors are running the operations. Even if major producers have source monitoring regimes that exclude using them, there are countless smaller producers that will save material costs by sourcing from uncertified supplies, thus maintaining the revenue flow that sustains destructive mining and violence.

The church's current efforts

The Catholic Church works in these processes in a number of ways. It accompanies affected communities, sharing in and documenting their suffering. This intimate knowledge forms the basis for policy advocacy. Advocacy on this level requires expert technical knowledge. Thus, ecclesial agency here tends to focus on elite participation: episcopal conferences and leaders and experts from their affiliated NGOs. Bishops call attention to destructive and exploitative practices and give voice to the plights of their victims. Episcopal conferences also provide a network which can monitor whether corporate and government action live up to promises made, as the Bishops of Cameroon and Chad did during construction of a pipeline in their region (US Congress 2007). Catholic Relief Services has worked to build capacity among local ecclesial and civil society groups so they can more effectively influence major extractive projects (Bamat, Chassy, and Warne 2011, 21). In El Salvador, the bishops worked with other Catholic and secular organizations in a successful campaign to pass national legislation to ban metal mining. Colombia is a more typical case. Multiple church actors are quite active in addressing mining issues at the local level. But these activities are not well coordinated and the bishops' conference has not developed a clear and consistent advocacy agenda at the national level. The chapter by Sandra Polanía-Reyes and Héctor Fabio Henao demonstrates how the Colombian Church is attempting to establish better coordination.

Because of their technical nature, such interventions are always susceptible to questions of religious legitimation. Do these appear to be part of the church's mission to people in the pews? They certainly are. But the abstract theological justifications for Catholic social doctrine often have little traction in the lived sacramental, liturgical, and communal experience of the Christian faithful. The technical nature of the regulation of extractive industries exacerbate this problem. For that reason, these sorts of actions often fail to gain strong support from much of the church and significant portions of the laity do not perform their expected role in this process of providing public pressure on governments to enforce environmental regulations, counter corruption, or ensure respect for the rights of threatened communities. This problem is why, as described by Polanía-Reyes and Henao, the bishops in Colombia have been trying to focus on better teaching and preaching of "eco-theology" that amplifies an understanding of the theology of creation in addition to a theology of salvation.

There is another problem inherent in the church's response, in addition to legitimation. It concerns the way the laity are addressed in such interventions, often as consumers. Greed and excessive consumption should, of course, be challenged. But invocations of greed can legitimate a false analysis that portrays market structures as morally neutral instruments that simply respond to the desires expressed in consumer demand.

Not only does this prevent us from critiquing the way our networks are structured, it also tends to refocus the ecclesial response to working on the demand side. Many important theological critiques (e.g., cultivation of concern for creation or solidarity with the oppressed) are offered in ways that implicitly address the laity as consumers and are thus construed as a means of moderating individual desire, which can distract attention from systemic causes woven into the fabric of industrial civilization and the global commodity market.

Pope Francis has articulated two important concepts that address structural dimensions of these problems. They are, however, often incorrectly interpreted in a way that reduces them to an individual level: "the throwaway culture" and "the globalization of indifference." The notion of a "throwaway culture" is an insightful critique of contemporary consumer-centered economies. Too often, however, this concept is reduced to an indictment of individual greed and consumer lifestyles. Consumer excess is a profound problem, but the crisis we face is not driven by the desires of the wealthy for shiny new gadgets. Yes, the tantalum and cobalt mined in conflict zones are used in the latest smartphones. But they are also in *every* electronic device—from hospital respirators to emissions-saving electric cars (Riofrancos 2019). As Daniel Finn (2017, 12) notes, extraction undergirds the whole of industrial society: "everything we touch throughout the day is either grown or extracted from the earth."

While Francis (2015b) does mention the need for a new "lifestyle," his extended discussion of the throwaway culture in *Laudato Si'* focuses on systemic issues of industrial production. He contrasts this with the circular flows of natural ecosystems, in which wastes from one part of the system feed and fertilize others. "Our industrial system … has not developed the capacity to absorb and reuse waste and by-products. We have not yet managed to adopt a circular model of production capable of preserving resources for present and future generations. …" (§16). Francis's critique of the throwaway culture involves much more than consumer excess. It cuts to the heart of industrial civilization.

Francis discusses the "throwaway culture" in the context of the "globalization of indifference" in *Laudato Si'*. Francis first spoke about this at Lampedusa in 2013. "In this globalized world, we have fallen into globalized indifference. We have become used to the suffering of others: it doesn't affect me; it doesn't concern me; it's none of my business!" (Francis 2013b). He developed this notion more fully in *Evangelii Gaudium*, where he placed indifference in a more systemic context. Francis (2013a, §53) called for expanding the Fifth Commandment—"Thou shalt not kill"—to the violence of economic systems: "we also have to say 'thou shalt not' to an economy of exclusion and inequality. Such an economy kills." He describes the globalization of indifference as a consequence of a view of the economy which assumes economic growth is adequate to bring about justice, trusting in the "sacralized workings of the prevailing economic system" (§54).

In Francis's fuller analysis, indifference is clearly characterized, not as an individual defect of will, but as a systemic disorder emerging from our economic context. My description of the three weaknesses of global market systems—externalities, distancing, and the attenuated form of relationships facilitated by the market's price mechanism—can be understood as an attempt to sketch the "architecture of indifference" in which the globalization of indifference emerges. It is not that "we should care, but don't." Rather, because of market externalities and commodity distancing "we don't know and can't know" the consequences connected to our consumption. Thus, it is impossible to care in any specific way (Miller 2014).

We have considered one of the structural factors that facilitates the destructiveness of mining: the way global capitalism connects the entire world in a network of shallow relationships, where the price signal is the primary form of communication that systematically obscures the broader conditions of production and destructive externalities. Laws, regulations, and industry standards that seek to counter these problems are always in danger of being outmaneuvered by market actors. We have seen that the church's involvement in these efforts suffers from two weaknesses. First, since it takes place on the level of experts and elites, it tends to be disconnected from the liturgical and sacramental life of the church. Second, in emphasizing the moral aspects of consumption, the church runs the risk of portraying these as demand-side problems arising from disordered desire, unwittingly obscuring the structural dimensions of the problem. In the final section we will explore how thinking in terms of networks will help re-envision the church's role in extractive markets and suggest new approaches for engagement.

A fuller ecclesial response

The church is materially, viscerally present on every level of this network. It is present in the communities who labor in mines and forests, who drink poisoned water, and who sometimes participate in the destruction of the land their children will inherit. It is present in the urban professionals supported by the proceeds of extractive industries and the wealthy parishes they attend. It is present in the consumers, rich and poor, who are fed, clothed, and housed by the products this system produces. The church is implicated at every level of these global chains, but those levels have little ecclesial relationship with one another. They are connected to one another, not by the church, but by the global commodity network. While theologically, we are united to one another in Christ, practically, we relate through global markets. The mystical Body of Christ is too often bound together by the sinews of global capitalism. The problem is thus not simply a matter of theology and morals, or of the formation of consciences. The problem is properly ecclesial, having to do with the nature and practice of the church. The church has something to say about the nature of human relationships. It must also consider the relationships in which it is entangled.

Henri de Lubac's classic text *Catholicism* aimed to show, in the words of its French subtitle, the "social aspects of doctrine." De Lubac (1988 [1938], 15) argued that Catholicism is social "in the deepest sense of the word." It is social "in the heart of its mystery, in the essence of its dogma" to such an extent that the phrase "social Catholicism" should sound redundant. As the means and fruit of God's work of salvation, the church is not only essentially social, but the ultimate form of human sociality which all humankind are destined to share in Christ. De Lubac did not view secular forms of community dialectically. He accepted their goodness, noting the ancient Christian use of the city as an analogy for the church. But he held that the church was always called to a deeper intimacy than any earthly community could provide. "Among those who are received within this heavenly city there is a more intimate relationship than subsists among the members of a human society, for among them there is not only outward harmony, but true unity" (113–15, quote at 115).

De Lubac's theology found expression in the Second Vatican Council's understanding of the church and its mission. *Lumen Gentium* defined the church as a "sacrament or ... sign and instrument" of "a very closely knit union with God and of the unity of the whole human race." Thus, the church's very reason for existing, its religious mission, is to bring humankind into union. The council went on to note that this mission is intertwined with the secular dynamics that were bringing about human interconnection. In its words, this sacramental working of the church brings a world "joined more closely today by various social, technical and cultural ties" to a fuller unity in Christ (Second Vatican Council 1964, §1).

The final document of the council, *Gaudium et Spes,* applied this understanding to the church's mission in the contemporary world. *Gaudium et Spes* spoke of the "socialization" being brought about by the growth of communications, and economic and political activity multiplying the "ties" among diverse men and women. This "growing interdependence" finds its fulfillment only on the "deeper level of interpersonal relationships." Such "communion between persons" fulfills the "laws of social life which the Creator has written into" human nature. *Gaudium et Spes* saw in the growth of global human interdependence the rise of a truly universal common good involving "rights and duties with respect to the whole human race" (Second Vatican Council 1965, §6, 23, 26).

One of the key terms for this social unity is "solidarity." *Gaudium et Spes* explicitly connected solidarity with the social nature of salvation beginning in the Jewish covenant: "From the beginning of salvation history He has chosen men [and women] not just as individuals but as members of a certain community." This continues in the "the work of Jesus Christ" who taught his disciples to "treat one another as brothers" and sisters. The church, "His Body," is a community where "everyone, as members one of the other, would render mutual service according to the different gifts bestowed on each." This ecclesial communion is solidarity, which "must be constantly increased until that day on which it will be brought to perfection. Then, saved by grace, men will offer flawless glory to God as a family beloved of God and of Christ their Brother" (§32).

The council understood that this communion has its ultimate realization as unity in Christ. My point here is not to stress that without Christ there can be no unity, but to note the very high stakes the church places on social communion. Thus, the shallow market relationships in which we are currently entangled are not secondary questions. They impinge upon the church's ability to pursue its mission to enact communion.

In *Laudato Si'*, Pope Francis developed the church's social teaching by expanding solidarity to all creation. Francis (2015b, §240) roots both solidarity among humans and the rest of creation in an ontology that reflects the triune, relational character of God.

> The divine Persons are subsistent relations, and the world, created according to the divine model, is a web of relationships. Creatures tend towards God, and in turn it is proper to every living being to tend towards other things, so that throughout the universe we can find any number of constant and secretly interwoven relationships. This leads us not only to marvel at the manifold connections existing among creatures, but also to discover a key to our own fulfilment. The human person grows more, matures more and is sanctified more to the extent that he or she enters into relationships, going out from themselves to live in communion with God, with others and with all creatures. In this way, they make their own that trinitarian dynamism which God imprinted in them when they were created. Everything is interconnected, and this invites us to develop a spirituality of that global solidarity which flows from the mystery of the Trinity.

Francis (2020, §22) developed these themes further in *Querida Amazonia*:

> Christ redeemed the whole person, and he wishes to restore in each of us the capacity to enter into relationship with others. The Gospel proposes the divine charity welling up in the heart of Christ and generating a pursuit of justice that is at once a hymn of fraternity and of solidarity, an impetus to the culture of encounter. The wisdom of the way of life of the original peoples—for all its limitations—encourages us to deepen this desire. In view of this, the bishops of Ecuador have appealed for "a new social and cultural system which privileges fraternal relations within a framework of acknowledgment and esteem for the different cultures and ecosystems, one capable of opposing every form of discrimination and oppression between human beings.

With these theological principles, we can judge the monstrous inadequacy of market relationships ruled by the price signal. But these insights do not simply judge an external reality. As we saw, the church is implicated in the structure of the global market. Its members relate to each other primarily through the market rather than through ecclesial connections. Thus, these theological principles are demands placed on the church itself to make such deep forms of relationship present in its own life.

Now that we appreciate the challenge that this global network poses to the church, we can begin to consider how it might respond. If the church is implicated in the destructiveness of global commodity flows and the shallowness of the human relations they construct, it can attempt to connect communities in a manner more consistent with its sacramental mission to unite humankind. If some members of the church are victims of conflicts fueled by mining, or fall ill from mining pollution, while others profit or consume products made from that mining, then the church can and should work to unite these members into deeper relationships of responsibility and reciprocity. Theologically, the church should aspire to be a network that encompasses and connects all its members, rather than merely be present in a network constituted by shallower logics. This task will always remain incomplete this side of the eschaton. But it is a work that, nevertheless, must be attempted.

There have been examples of success. Sharon Erickson Nepstad's account of the Central American Solidarity movement in the 1980s provides a hopeful and illuminative example. In a context riven by national boundaries and Cold War geopolitics, US and Central American Christians found ways to work together in solidarity to address the needs of those caught in the violent civil wars of that era. Nepstad recounts how North American missionaries and parish relationships served as an ecclesial connection across state borders. Returning missionaries served as bridge figures in their home communities, testifying convincingly to the innocence of victims of political violence at a time when the US government portrayed them as dangerous radicals (Nepstad 2004, 67–75). In these contexts, the oppressed appeared not as foreigners, but as brothers and sisters in Christ. Responding to their needs seemed, therefore, more than a matter of cosmopolitan ethics, but a question of Christian discipleship. It was a significant struggle for North American Christians to question their government and organize to change US foreign policy. The communications network provided by missionaries, intra-church partnerships, and church media provided knowledge and relationships that motivated and sustained advocacy by North American laity on behalf of their brothers and sisters in Central America.

The key lesson of this example is that the church provided alternative communication networks to government and secular media that enabled more direct communication between its members. Solidarity arose from this communication. It is important to note how this communication occurred in an ecclesial setting where it was easily connected to the liturgical and sacramental beliefs and practices of the church.

How might something like this be enacted in the context of global commodity chains? It would involve cultivating properly ecclesial relationships to deepen the hidden, shallow relationships that currently connect us through commodity mining markets. The Catholic Church is one of the few organizations that has as broad a reach as the globalized commodity network. The church currently has several communication networks that are relevant. None, however, are used for this task.

The most central network of the Catholic Church is its episcopal hierarchy, which is in both theological communion and bureaucratic connection with the Vatican. While these networks are the most often invoked in discussions of catholicity, they tend to function primarily in spoke-like fashion, connecting each local church to the Church in Rome. They are less effective in connecting the local churches to one another. National and regional episcopal conferences allow for regional communication. New structures are needed to allow point-to-point global communication among local churches on a global scale. The planned continent-level discussions in the 2021–2023 Synod on Synodality have potential to address this problem.

The church also has the extensive networks of its Caritas affiliate organizations that connect the churches in the Global North with those in the Global South. These are impressive networks that can provide relief, development aid, and conflict resolution resources to communities in need (regardless of religious affiliation). Although they are not designed to establish communication between different parts of the church, some do fulfill this role. Catholic Relief Services, for example, facilitates global advocacy with its Catholics Confront Global Poverty partnership with other entities of the United States Conference of Catholic Bishops, and it organizes speaker programs to bring representatives from the field into the United States. This, however, takes place as a secondary supplement to its massive Overseas Operations which handles its disaster, development, and peacebuilding work. These are networks that provide resources and expertise to struggling communities. What if they were developed to allow for two-way communication that could allow needy communities to not only receive assistance, but to speak their experience and insights to the rest of the church?

Pope Francis (2015a) has sought to renew the Synod of Bishops in order to facilitate a "Church which listens." The 2019 Synod for the Pan-Amazonian Region provided a noteworthy advance in the church's work of advancing global solidarity focused on developing communication from the peripheries of the church and the global market. The preparation for the Synod involved hundreds of listening events in which more than 87,000 people participated (Brockhaus 2019). The working document for the Synod called for a "A dialogue in favour of life…at the service of the 'future of our planet,' of transforming narrow mentalities, of converting hardened hearts, and of sharing truths with all humanity" (Synod of Bishops Special Assembly for the Pan-Amazonian Region 2019, §40).

In his response to the synod, Francis spoke of this listening in terms evocative of the catholicity of the church. We listen not simply to hear other's struggles, but to learn what each culture has learned of God's wisdom in its landscape and ecology.

> Human groupings, their lifestyles and their worldviews, are as varied as the land itself, since they have had to adapt themselves to geography and its possibilities.... In each land and its features, God manifests himself and reflects something of his inexhaustible beauty. Each distinct group, then, in a vital synthesis with its surroundings, develops its own form of wisdom. (Francis 2020, §32)[5]

The Synod on the Pan-Amazonian Region provides a glimpse of how ecclesial networks can provide a deeper connection between peoples than the global market system (Miller 2020). People around the world consume goods originating in the Amazon—from gold and iron to palm oil and soy—yet there are few opportunities to hear from those affected by their production in any depth. For three weeks in October of 2019, the church sought to "listen to the voice of the Amazon, stirred by the powerful breath of Holy Spirit in the cry of the wounded land and its inhabitants" (Synod of Bishops Special Assembly for the Pan-Amazonian Region 2019, §3).

Evaluating mining projects requires an enormous depth of knowledge: the nature of the mining and refining processes being used, the ecology of the site, the economic, social, and cultural situation of those impacted by the project, the politics of the local concessions process, and the seriousness of negotiations with affected parties. While it is true, as Daniel Finn notes, that one can easily find blanket condemnations of all extractive projects that ignore the significant efforts some extractive corporations make to work in a just and sustainable manner, it is also true that for just about any project, even notoriously destructive ones, one can find legitimate-sounding documentary evidence of consultation processes, negotiations, and compensation. Only if one has access to detailed local knowledge can this evidence be judged adequately.

The Catholic Church has profound resources that could be mobilized on these issues. It is often present in the communities affected by mining projects and can help them voice their concerns. It can provide expert technical support and connect them with national, regional, and international networks. Rigobert Minani's discussion in this volume provides the compelling example of the work of the Episcopal Commission for Natural Resources in the Democratic Republic of the Congo. It has established diocesan "natural resource observatories" that monitor extractive projects for human rights violations. It works on the national level to advocate for just laws and policies for affected communities, and recently ran a program to help enfranchise artisanal miners so they can better navigate the legal systems for transparency monitoring and better participate in the marketplace. Working with civil society groups, it connects local concerns to advocacy on national, regional, and international levels.

Andrés McKinley's analysis in his chapter of the church's role in El Salvador's successful campaign to end metallic mining displays similar dynamics. The church, as McKinley describes, addressed the vulnerability and isolation of poor communities by providing the means to connect their resistance with advocacy campaigns on the national level. Church actors also facilitated a connection between global communities affected by the same mining firm by sponsoring the visit of a governor from a community in the Philippines to share their experience of the mining company's actions. In both cases, the church served as a network to connect communities that would otherwise be isolated so that they could better respond to the challenge posed by powerful mining interests. This connection between El Salvador and the Philippines was made through an international NGO that works on mining issues. There is nothing wrong with that. Note, however, the unrealized potential of the church's own network which could have been used to connect these communities in a more ecclesially powerful manner.

Guided by the principle of synodality, the church should further develop networks that enable people in extractive zones to speak with their brothers and sisters around the world; to share both their understanding and predicaments. If the contextual facts of every mining project are narratively dense, the church could provide connections that allow the richness of the stories to be shared: the impacts—for good and ill—on local populations; the landscapes and ecologies that are affected; and equally importantly, the local culture's understanding of the meaning of the landscapes they inhabit and the ecologies in which they participate. This could facilitate, in Leah Riofrancos's (2019, n.p.) words, "broadening our solidarities to encompass people we may never meet and places we may never visit but whose futures are bound up with our own."

If such communication were fostered by the church, it would make the religious relevance of these issues much more apparent. Rather than reading reports from experts (in our age of skepticism), believers could listen to the experience of fellow parishioners from across the globe. This would deepen their sense of connection and obligation. Equally important, by encountering the people at the other end of the supply chain, it would reveal to them how much is ordinarily hidden by our global supply chains. Knowing those affected by mining could foster a deeper inquisitiveness and distrust of products which do not disclose their supply chains (see Miller 2004, 184–86).

Such synodal listening to those at the extractive ends of global commodity chains could radically deepen the practice of FPIC, which currently functions largely as a contractual formality, inviting affected communities into a due-diligence-like review of a proposed project. In its most sincere forms, it allows local communities to refuse a project. But it primarily imagines negotiations to minimize harms and find meaningful compensation. Francis's association of synodality with catholicity opens up the possibility for a deeper dialogue. Local communities could be given more than a "take it or leave it" offer for extractive development. A synodal communication network would allow them to dialogue with industrialized cultures about the nature of civilization. In the Amazon, there are indigenous civilizations that have lived in sustainable harmony with their ecologies for more than ten thousand years. We in industrialized societies, who have in a few short centuries mined our minerals, felled our forests, and now face ecological collapse on multiple fronts, have much to learn from such cultures. Rather than limiting communication to seeking their consent to our mining projects, the church could foster a deeper communion that would allow a two-way dialogue between parties already tied together via commodity flows. Listening would first confront us with the question, "Is our way of life worth the destruction of theirs?" Such a dialogue would also call industrialized society's way of life into question, helping to transform a civilization that is currently transgressing planetary boundaries on so many fronts.

This synodal ecclesial communications network does not currently exist. From a historical perspective, this is nothing new. The networks of Catholic communion are constantly changing and developing: from the Tridentine adoption of the printing press to foster liturgical and theological uniformity to the development of episcopal conferences to better organize church responses to the modern nation-state. The technology exists to enable the church to deepen its communications to match this theological need. If social media can sustain debate and friendship across space, they can certainly connect believers among distant local churches. If tens of millions daily watch livestreams of videogames and e-sports, these same technologies can certainly facilitate conversations between the peripheries and centers that can deepen relationships beyond those fostered by global capitalism.

Conclusion

This chapter has argued that many of the destructive impacts of mining are facilitated by the defects and cultural impacts of global commodities markets. The Catholic Church is itself a network, one that aims for a profound communion among all people. The fact that its members are currently more strongly formed by market relations than by its own logic of communion is a profound challenge. This challenge provides the possibility of new response, supplementing its current work of technical advocacy. By seeking to build ecclesial connections among its members, the church could form synodal relationships that deepen current market relationships into fuller responsibility and allow those affected to become full partners in a dialogue about our shared future in our common home.

Notes

1 The chapter in this volume by Sandra Polanía-Reyes and Héctor Fabio Henao describes these different scales in the context of Colombia.
2 Also see Finn's more extended account of the morality of participating in global markets in his 2019 book *Consumer Ethics in a Global Society: How Buying Here Causes Injustice There.*
3 There is significant debate about the effectiveness of the Dodd-Frank provisions concerning conflict minerals. See Koch and Kinsbergen (2018) and Diemel and Hilhorst (2019).
4 For a discussion of the inadequacy of FPIC see Thea Riofrancos (2020). On the "Right to say no," see Justice, Peace, and Integrity of Creation Commission of the Union of International Superiors General-Union of Superiors General (2019), which proposed fostering the right of communities to say "no" to mining as a focal point of these religious orders' future work on the issue.
5 See also Francis's (2018) address at a meeting with indigenous people at Puerto Maldonado, Peru.

References

Arboleda, Martin. 2020. *Planetary Mine: Territories of Extraction Under Late Capitalism*. London: Verso.

Bamat, Tom, Aaron Chassy, and Rees Warne, eds. 2011. *Extractives and Equity: An Introductory Overview and Case Studies from Peru, Angola and Nigeria.* Baltimore, MD: Catholic Relief Services. https://www.crs.org/our-work-overseas/research-publications/extractives-and-equity.

Brockhaus, Hannah. 2019. "The Amazon Synod, by the Numbers." *Catholic News Service*, October 11, 2019. https://www.catholicnewsagency.com/news/the-amazon-synod-by-the-numbers-11205.

Castillo, Daniel P. 2019. *An Ecological Theology of Liberation: Salvation and Political Ecology*. Maryknoll, NY: Orbis Books.

de las Casas, Bartolomé. 1986. *Historia de las Indias*. Vol. 2. Edited by André Saint-Lu. Caracas: Biblioteca Ayacucho.

de Lubac, Henri. 1988 [1938]. *Catholicism: Christ and the Common Destiny of Man*. Ignatius Press: San Francisco.

Diemel, J.A., and D.J.M. Hilhorst. 2019. "Unintended Consequences or Ambivalent Policy Objectives? Conflict Minerals and Mining Reform in the Democratic Republic of Congo." *Development Policy Review* 37, no. 4: 453–469. 10.1111/dpr.12372.

Finn, Daniel. 2017. "Recycling isn't enough: A Defense of Responsible Mining." *Commonweal*, May 2, 2017. https://www.commonwealmagazine.org/recycling-isn't-enough.

Finn, Daniel. 2019. *Consumer Ethics in a Global Society: How Buying Here Causes Injustice There*. Washington, DC: Georgetown University Press, 2019. 10.2307/j.ctvswx7rz.

Francis. 2013a. *Evangelii Gaudium*. http://www.vatican.va/content/francesco/en/apost_exhortations/documents/papa-francesco_esortazione-ap_20131124_evangelii-gaudium.html.

Francis. 2013b. Homily in "Arena" sports camp, Salina Quarter, Lampedusa. July 8, 2013. http://www.vatican.va/content/francesco/en/homilies/2013/documents/papa-francesco_20130708_omelia-lampedusa.html.

Francis. 2015a. "Address Commemorating the 50th Anniversary of the Institution of the Synod of Bishops." October, 17 2015. http://w2.vatican.va/content/francesco/en/speeches/2015/october/documents/papa-francesco_20151017_50-anniversario-sinodo.html.

Francis. 2015b. *Laudato Si'*. http://www.vatican.va/content/francesco/en/encyclicals/documents/papa-francesco_20150524_enciclica-laudato-si.html.

Francis. 2018. "Address to Meeting with Indigenous People of Amazonia, Coliseo Regional Madre de Dios (Puerto Maldonado)." January 19, 2018. http://www.vatican.va/content/francesco/en/speeches/2018/january/documents/papa-francesco_20180119_peru-puertomaldonado-popoliamazzonia.html.

Francis. 2020. *Querida Amazonia*. http://www.vatican.va/content/francesco/en/apost_exhortations/documents/papa-francesco_esortazione-ap_20200202_querida-amazonia.html.

Galeano, Eduardo. 1997. *Open Veins of Latin America: Five Centuries of Pillage of a Continent*. New York: Monthly Review Press.

Justice, Peace, and Integrity of Creation Commission of the Union of International Superiors General-Union of Superiors General. 2019. "Statement from the Participants of the JPIC & Mining Seminar." September 27, 2019. https://www.jpicroma.org/single-post/2019/08/27/JPIC-MINING-WORKSHOP-September-25–27.

Koch, Dirk-Jan, and Sara Kinsbergen. 2018. "Exaggerating Unintended Effects? Competing Narratives on the Impact of Conflict Minerals Regulation." *Resources Policy* 57: 255–263. 10.1016/j.resourpol.2018.03.011.

Lujala, Päivi. 2018. "An Analysis of the Extractive Industry Transparency Initiative Implementation Process." *World Development* 107: 358–381. 10.1016/j.worlddev.2018.02.030.

Miller, Vincent J. 2004. *Consuming Religion: Christian Faith and Practice in a Consumer Culture*. New York: Continuum.

Miller, Vincent J. 2014. "Slavery and Commodity Chains: Fighting the Globalization of Indifference." *America Magazine*, January 3, 2014. https://www.americamagazine.org/issue/slavery-and-commodity-chains-fighting-globalization-indifference.

Miller, Vincent J. 2020. "We're All Tied to the Amazon Through Globalization. Can the Synod Help Us Listen to Its Cry?" *America Magazine*, February 19, 2020. https://www.americamagazine.org/faith/2020/02/19/were-all-tied-amazon-through-globalization-can-synod-help-us-listen-its-cry.

Nepstad, Sharon Erickson. 2004. *Convictions of the Soul: Religion, Culture and Agency in the Central American Solidarity Movement*. Oxford: Oxford University Press. 10.1093/0195169239.001.0001.

Princen, Thomas. 2002. "Distancing: Consumption and the Severing of Feedback." In *Confronting Consumption*, edited byThomas Princen, Michael Maniates, and Ken Conca, 103–132. Cambridge, MA: MIT Press. 10.7551/mitpress/2097.001.0001.

Riofrancos, Thea. 2019. "What Green Costs." *Logic* 9. https://logicmag.io/nature/what-green-costs.

Riofrancos, Thea. 2020. *Resource Radicals: From Petro-Nationalism to Post-Extractivism in Ecuador*. Durham, NC: Duke University Press. 10.2307/j.ctv14t48kg.

Second Vatican Council. 1964. *Lumen Gentium*. http://www.vatican.va/archive/hist_councils/ii_vatican_council/documents/vat-ii_const_19641121_lumen-gentium_en.html.

Second Vatican Council. 1965. *Gaudium et Spes*. https://www.vatican.va/archive/hist_councils/ii_vatican_council/documents/vat-ii_const_19651207_gaudium-et-spes_en.html.

Synod of Bishops Special Assembly for the Pan-Amazonian Region. 2019. *The Amazon: New Paths for the Church and for an Integral Ecology*. http://www.vatican.va/roman_curia/synod/documents/rc_synod_doc_20191026_sinodo-amazzonia_en.html.

UN General Assembly. 2007. *UN Declaration on the Rights of Indigenous Peoples*. A/RES/61/295. September 13, 2007. https://undocs.org/A/RES/61/295.

US Congress. 2007. House. Committee on Financial Services. *Transparency of Extractive Industries, High Stakes for Resource-rich Countries, Citizens, and International Business. 110th Congress*, First Session, October 25, 2007. Statement of Father Patrick Lafon, former Secretary General of the Catholic Bishops' Conference of Cameroon. https://www.govinfo.gov/content/pkg/CHRG-110hhrg39913/html/CHRG-110hhrg39913.htm.

15 The mining industry: the journey from impunity to consent

Raymond Offenheiser

For millennia, humans have valued metals and precious stones, committed heinous crimes to obtain them, traveled great distances to trade them, and created and sustained empires based on their real or imputed value. Spanish conquistadors took an Incan emperor hostage so he would reveal gold sources. Many lives over centuries have been sacrificed at mines, often brutally and without accountability. Until the late twentieth century, mining required large numbers of laborers working at great risk for paltry wages to extract precious metals and ores.

Happily, this is not the world we live in today. As one mining executive remarked to me casually: "There is no backyard on earth today where we can dig a hole without the world knowing about it in 24 hours." This is not to say there are no longer any bad actors, horrific working conditions, or human rights violations. It's still hard to find any place on earth where there is not some conflict over mining. Moreover, we live in a globalized world with exploding demand for minerals. It's estimated that some ninety metal and mineral commodities are required to power the global economy (US Geological Survey n.d.).

Those interested in engaging mining companies in dialogue and driving significant change need to recognize that a lot of normative progress has been made in a relatively short period of time. Today there are dozens of voluntary normative frameworks focused on ethical issues, human rights, specific commodities, industry operational practices (tailings dams), transparency and taxation, and certification standards (World Economic Forum and Global Agenda Council on the Future of Metals and Mining 2015). These new frameworks influence the industry and create tools for Catholic organizations and others to scrutinize and propose changes to industry policies and practices, and to advance bold regulatory reform.

While progress in realizing rights and justice on some fronts has been made, the normative frameworks for extraction need to be refined and taken from C-suites to mine sites, through supply chains for full implementation, and, finally, to legislatures. Catholic and other civil society actors should build upon these norms and incorporate them into their

DOI: 10.4324/9781003094272-15

strategies as they push the industry toward greater sustainability, peace, and integral human development.

Ethics and mining

Mining presents a vast array of ethical questions at multiple levels, but the idea that there should be an ethics for mining is relatively new. While norms for agriculture, forestry, and fisheries have been included in international environmental and development summits and natural resource policy documents since the early 1970s, mining was somehow overlooked. Not until the late 1990s was consideration given to the environmental, and later, the social, and economic impacts of mining. Mining is foundational to modern life and, as George Orwell described it, part of the "metabolism" of civilization. Yet there has been little scrutiny until recently of the complex issues of conflict, poverty, and injustice that surround it, to say nothing of its growing strategic importance and long-term sustainability.

In *Laudato Si'*, Pope Francis examines the moral challenges posed by mining on a planet with limited natural resources and under pressure from a global climate crisis. He offers considerations for how unfettered growth and a global consumer culture affect the calculus of costs and benefits in industries like mining. He questions how much the prevailing consumerist culture that depends on mining is itself in need of drastic change. In his development of the idea of integral ecology, he delivers a mandate for Catholic leaders and activists to tie their work to the machinations of the globalized economy, the consumerism that drives it, and its immense consequences for "our common home." Mining is squarely within that purview.

Despite these ethical concerns, can we live without mining metals and minerals or are they essential for the survival and advancement of civilization? Can we live without steel girders that support our modern buildings, bridges, and infrastructure, or lithium for batteries that electrify our homes, or copper that supports our electrical grids? There are 62 unique and irreplaceable metals in every cellphone. Each one must be mined somewhere at some cost to environment and society. Is securing those metals a cost we can and must bear? While the mining industry is exploring the development of synthetic minerals and mining the moon, it is unlikely these efforts will be sufficient to address the exploding demand for critical minerals needed for information technology, mass communication, and energy any time soon.

There is certainly a case to be made for challenging extractive industries and their wholesale and retail customers to embrace responsible stewardship of the planet and envision a world in which extraction is limited to the essential, but how do we define essential? Do we devalue precious stones and jewelry? Do we impose regulatory limitations mining low-grade mineral deposits for certain precious stones and metals? Do we increase investment in the search for synthetic minerals that would reduce environmentally destructive extraction? Such efforts would require major changes in public taste and values but are goals that should be under consideration. These kinds of questions point toward the complexities that lie before us in trying to limit and regulate the scale and impact of the mining industry. We need it, yet it will continue to be a dirty business riddled with ethical dilemmas.

But to tackle these ethical questions and be effective in addressing them, Catholic peace, human rights, and development activists will need to become more knowledgeable about the inner workings of this industry: its business ecosystem, its profitability, its strategic priorities, its internal incentives and policies, its technical complexities, and its financial and operational vulnerabilities. They will need to also reckon with the unique contributions the mining industry makes to society as they respond to the pope's critique of consumerist culture and an economy that prioritizes growth over other values.

Understanding the mining mindset

In seeking to transform the mining industry, it is critical to understand its worldview, values, and culture. While mining technology has improved over the centuries, it is still fundamentally about digging large holes in the earth to extract value from a very small quantity of minerals. Today, ore grades have declined globally, so companies must move ever-increasing amounts of earth to extract very little real value. One gold wedding band may require the removal and processing of thirty tons of earth.

The industry is largely led by mining engineers and geologists who are trained to determine the smartest way to extract the maximum amount of mineral value from a discrete piece of ground. They are single-minded about this task. What matters to them is what is in the ground inside the fence where they operate. They are linear thinkers and they have little regard for how what they are doing inside the fence might affect those living outside the fence. These geologists and engineers define the culture of the mining industry. Today, however, this culture is under increasing pressure to change in fundamental ways.

For many centuries, mining was able to exploit surplus land and labor with little regard for safety, human rights, wages, or ecological or community impacts. But companies can no longer treat environmental damage, injury, or death as incidental costs of doing business. Having suffered major legal and financial setbacks over the last fifty years, companies are highly sensitive and proactive about safe working conditions. Serious environmental catastrophes like the Samarco tailings dam collapse in Brazil in 2015 have demonstrated to executives the scale of liability possible from lax management of critical risks. In perhaps an equally dramatic example, as detailed by Andres McKinley in this book, the perceived downstream risks of water shortages to agriculturalists led El Salvador's Congress to ban all large mining projects in the country, the first nation in the world to take such drastic legal action. Events like these and increasing investor scrutiny of environmental, social, and governance policies and practices have forced particularly the largest companies to rethink their business models and incorporate new ethical and performance policies. These new standards represent a real shift by companies to incorporate ethics guidelines for what they will commit to do "outside the fence."

The industry's first efforts were to embrace broad principles of corporate social responsibility while treating challenges as a public relations matter. They later reluctantly embraced the concept of social license as a recognition that public relations would not suffice and that they had to engage communities in new ways. The initial ways in which companies tried to acquire social license were perceived by communities and civil society as short-term transactional approaches that did not address long-term development interests and justice concerns. But in the late 1990s, leading companies, or Majors,[1] began to engage a broad spectrum of non-industry stakeholders. In the early 2000s, the Majors created the International Council on Mining and Metals (ICMM) as a joint initiative to address a wide array of shared concerns, such as sustainability, indigenous rights, waste management, and mine closure. To join, companies had to pledge to implement the ICMM Ten Principles, which give industry guidance across ten areas of ethical consideration: ethical corporate governance; sustainability-driven decision-making; human rights; stakeholder-sensitive risk management; worker health and safety; environmental performance; conservation of biodiversity; support for responsible metal and mineral disposal, reuse, and recycling; social performance; and stakeholder engagement (International Council of Mining and Metals n.d.a). As time has gone on, expectations have been raised and additional dimensions like gender and equitable distribution of costs and benefits have been included.

Mining companies have to modernize, not only technologically but in terms of values and approach. They need to hire outside their traditional circles of geologists and engineers and incorporate personnel who bring ethical, environmental, economic, social, and political perspectives aligned with sustainability and human rights. At the same time, the promise of the reforms achieved thus far will not be fully realized without significant and continuing pressure from key investors, consumers, peacebuilders, and human rights activists. The moral authority and influence of the Catholic Church give it a prominent role in this regard.

A critical perspective on normative progress

It is important to recognize that most of what has been achieved is in the form of ethical frameworks and voluntary principles, or soft law. Very little has been achieved in terms of transforming soft law into hard law—i.e., legislation and regulation. The consequence of this is uneven performance from the industry, and likely the need over time to transition some voluntary standards to hard law. Industry leaders understand this and are the first movers in adoption and implementation. They would prefer to move quickly on controversial issues and stay ahead of aggressive regulatory campaigns where they feel voluntary principles would both suffice and afford them greater flexibility in their responses. Second-tier companies, or Juniors, have been less responsive than the Majors, who are all ICMM members.

While not always the ideal outcome for mining reform advocates, this proliferation of normative frameworks is generating policy debate within corporate board rooms and trade associations, which represents enormous progress over where things were even twenty years ago. It is useful to critically review progress across different areas that have been sources of conflict.

Environmental impact

Environmental impacts are one of the oldest and most common sources of conflict between companies and communities. Hence, it is no surprise that environmental impact has been the area on which companies have moved most readily. There are several reasons for this. While environmentalism began earlier in industrialized countries, by the early 1990s environmental movements were growing in countries across the world. The Rio Climate Summit brought together these national movements and enabled them to forge common bonds. Governments were soon establishing environmental ministries and requiring environmental impact assessments, and investor groups were taking note and incorporating environmental principles into their assessment criteria. Another reason for the readiness of companies to respond to environmental concerns is to avoid expensive legal fights and reputational damage. The famous case of the Texaco oil spill in the Ecuadorian Amazon that dragged on over two decades in US and Ecuadorian courts is well-known within the industry. There have been other cases as well, perhaps less publicized, but no less costly.

Companies have gotten very sophisticated about their environmental management capabilities and have incorporated environmental engineers into their teams to monitor a variety of potential impacts, including chemical pollutants, dust, tailings spills, drainage, water quality, and carbon footprint. In the most sophisticated cases, they are even monitoring regional biodiversity and conservation far beyond their mine sites. Given the scientific base for environmental monitoring, it has been relatively easy to fit environmental scientists into their existing staff configurations. The larger challenge has been to get company culture to recognize that there may be situations when environmental conditions are not favorable to extraction due, for example, to regional watersheds or sub-surface water systems.

At the country level, environmental impact assessments are done by the interested company and not by independent bodies. Depending on the national context, companies may understate risks knowing that governments will not conduct secondary reviews. In other cases, governments wanting the revenue will simply overlook any perceived risk and then not enforce or penalize companies when there are significant impacts on communities and public health. Environmental impacts are often felt from the earliest days of a project and for generations beyond its closure. While progress has been made in the upfront assessment of environmental risk, there is still a great deal to be done by both governments and companies in determining who will pay for downstream impacts over the duration of a project and beyond. The most sophisticated companies are doing lifecycle cost assessments and assuming in perpetuity liability for certain kinds of projects. But many governments and community groups lack the sophistication to do this kind of lifecycle assessment, build appropriate costs and conditions into project approvals and long-term contracts, or pressure companies not acting with good will to do so.

Companies and governments have made significant progress on environmental concerns, but performance across the industry is still uneven and requires independent monitoring. The industry will improve its performance only in response to citizen action and more effective government regulation and enforcement. Advocates interested in integral human development and peacebuilding would be wise to engage with communities, governments, and environmentalists in the early stages of project development in order to address the knowledge asymmetry and bring the best science and negotiating skills to contract negotiations with companies on behalf of affected communities.

Indigenous rights and other human rights

Human rights is an area fraught with the potential for violent conflict and requires ongoing oversight and vigilance. Unlike with environmental issues, mining companies lack competency in this field and yet recognize it is an area of considerable risk and vulnerability.

As illustrated in the chapter on Peru by José Bayardo Chata Pacoricona, indigenous rights violations continue to be one of the principal areas of contention and conflict between companies and communities. Many mining projects are located on lands titled under legacy treaty arrangements to indigenous populations. They may be aboriginal peoples in Australia, low caste tribal groups in India, or native populations in the depths of the Amazon. Quite often, these lands have powerful cultural importance. In many cases, governments have ceded indigenous communities the right to surface lands while retaining the rights to the sub-soil. Conflicts ensue when the state grants exploitation rights to extractive industries that then must open vast tracts of surface land to access their claims and begin extraction.

Very often, these deals are consummated without any consultation with the indigenous population. Once the first access road is opened up, colonists quickly follow, staking claims to indigenous territory. Governments simply look the other way and ignore any claims by indigenous communities to protect their territorial rights.

In the past, governments would simply force access for the companies with police action. Today, this kind of frontal assault on the rights of indigenous people would be actively resisted, leading to violence and likely the killing of indigenous peoples. Companies are now taking entirely different approaches. In Canada, where the government has strengthened the rights of First Nations over their traditional lands, tribes are negotiating equity partnerships over mining properties. In Australia, companies are negotiating collaborative agreements with aboriginal owners over access, water management, and, even in some cases, restoration and management of traditional lands. BHP, the largest mining company in the world, recently announced its support for proposals to amend the Australian constitution to recognize the territorial rights of traditional owners (Australian Associated Press 2019).

While these are welcome steps toward wider recognition of indigenous rights, there are many other places where indigenous rights are still widely abused. Governments often find it challenging to balance the collective ownership rights and cultural practices of native peoples with their desire to promote the economic ambitions of a major corporation. In Peru, a recent administration sought to pass legislation undermining the concept of collective title. In African nations, collective lands are often under the control of tribal chiefs who will bow to government pressure and make deals with private companies with no consultation with their tribal constituency. A government's lack of respect for collective ownership either undermines the entire historic titling of lands to native peoples or forces a tribe to parcel out lands to their members so that the individuals might more easily be forced to forfeit ownership or sell to interested commercial entities. While this would all be technically legal, we are likely to continue to see human rights challenges where indigenous collective rights of ownership are at odds with the capitalist impulse to commodify land in service to economic interests.

Apart from conflicts over indigenous rights, it is not uncommon for local communities to experience violent confrontations with security services contracted by companies to patrol and defend the company's property boundaries. And it is also not uncommon for those security forces to act with impunity due to the backing of government powers. A representative case comes from a nickel mine outside El Estor, Izabal, Guatemala. The mine was operated by the Canadian firm Hudbay Minerals. The mine's security forces perpetrated years of violence against the surrounding community, including assassination, rape, forced displacement, and criminalization of protest. In 2009, community leader Adolfo Ich was killed, setting off a series of court cases in Guatemala and

Canada (Russell 2019). The chief of security who killed Ich was finally convicted in January of 2021.

When confronted with violations of human rights or financial malfeasance, companies prefer a soft-law solution in the form of voluntary principles, such as the widely-accepted Voluntary Principles on Security and Human Rights (Voluntary Principles Initiative n.d.). Such a solution shields companies from real liability and leaves room for them to manipulate how their performance is gauged. This is not to say that voluntary frameworks like these are of little value. Their mere existence forces companies to raise human rights to a top-level priority with their operational managers, challenging them to carefully screen and train all security personnel to ensure compliance with human rights principles. Some companies now routinely hire independent human rights lawyers to evaluate their human rights risks and recommend policy and practice reforms. In one notable case, Newmont Mining walked away from its Conga Project in Cajamarca, Peru, a multi-billion-dollar gold and copper mine, after various violent incidents and protracted conflicts with local citizen groups and the regional government.

Companies like Newmont recognize that absent community consent, they risk committing serious human rights violations and then facing lengthy prosecution and costly court rulings. Companies have come to realize that they must find new and constructive ways to engage. Free, prior, and informed consent and social license have accordingly become mainstreamed parts of most mining companies' basic operations, and they are perhaps the most powerful normative tools available to defend indigenous land claims and the rights of local communities.

Free, prior, and informed consent and social license to operate

Free, prior, and informed consent (FPIC) is a response to companies securing mining rights from governments with little or no consultation with local communities. The goal of FPIC is to give a community the opportunity to conduct a public referendum on whether it endorses proposed mining in its region.

The first public use of this concept was in Tambogrande, Peru, a small town in a river valley on the Peruvian north coast whose economy is entirely dependent on irrigated agriculture. The community was concerned that mining operations would consume too much water upstream and undermine their agriculture downstream. The town organized a referendum to test public sentiment and the vote was 96% against the mine. The Peruvian government accepted the community's decision and Manhattan Minerals had to abandon the project. This decision had a profound effect on the mining industry. Prior to that case, industry presumed that a license to explore was implicitly a license to exploit. After the Peru decision, social license was no longer an abstract concept, but something requiring formal approval by host communities. This was an unexpected and revolutionary turn of events.

There have been few other referenda like the one in Tambogrande, but companies are wary and on notice. Their response has been varied. Some companies, largely Majors, have in principle incorporated the idea of FPIC in their policies and public statements. Others have not gone that far. Even in cases when the community right to FPIC is acknowledged, the company preference is to interpret the process as free, prior, and informed *consultation* rather than *consent*, in keeping with the accepted language of the ICMM (n.d.b). That word change can prevent FPIC from having real force, keeping communities from having the right to say "no" to a project and making the consultation frequently amount to nothing more than showy formality. The notion that communities can undermine substantial corporate investment through a community-led referendum is still an idea most companies are not ready to fully endorse. Companies would prefer to avoid a referendum and do their own assessment of community support before ever putting a shovel in the ground. Especially when backed by governments, FPIC has clearly given communities a powerful normative tool that companies feel obligated to observe. And while FPIC is still often undermined or watered down, it has done enough to make the industry shift to a more robust focus on acquiring genuine social license to operate in local communities.

Securing social license to operate is one of the most important issues for company executives today. When a company invests large tranches of capital into a particular project, it needs to exploit that mine for as much as thirty years before it can realize a significant profit. Companies used to operate with little concern for the impact on the wider community. However, the Minahasa Raya mine experience in Indonesia, in which Newmont Mining had multiple executives jailed on charges of dumping tailings into a nearby bay, and the Cajamarca case of mercury poisoning in a rural Peruvian village taught companies that such incidents seriously undermine public trust and can lead to complex problems that stall production for months.

In its earliest use, social license posed the simple question to company personnel: What is it going to cost to buy off community dissent so that we can proceed with operations? Over time, companies realized that their approach to community consultation was flawed in several respects. The personnel charged with managing community relations were simply not trained or equipped to take on this role. They did not understand complex community dynamics; were impatient and looking for a quick fix; and were poor tacticians and negotiators. They generally favored a very narrow and advantaged segment of the community and excluded other, more critical voices. They tried to offer small contributions for the construction or improvement of a school or clinic that was disproportionate to the scale of need, expectations of communities, and profit realized by the company. Many company leaders recognized that they needed to recruit and develop new personnel with the skills and competencies to work successfully outside the fence.

The pursuit of social license, with all the flaws described, remains the strategy of many companies. However, others have gone further and embraced the idea of "shared social value" (Porter and Kramer 2001). Shared social value assumes that the values of the company and society should align. With every new project, a company should appreciate the broad societal value it might bring to the communities directly or indirectly impacted by its operations. Discussions on social license should take place throughout the entire lifecycle of a mine. On its face, shared social value is potentially a dramatic leap forward from the highly transactional, myopic, one-and-done, and self-serving notion of social license and the preferred soft consultative version of FPIC. In theory, it signals a company's embrace of a larger regional development vision in its own planning and investments. In practice, this could mean engaging in elaborate multi-stakeholder planning for how mining operations contribute to regional development priorities. It could mean mine lifecycle planning with local and regional communities to reach some consensus on a post-extraction vision for the region and its resources. It could mean sizeable investments in regional infrastructure and services outside the confines of the mine. It could mean joining with local public officials to advocate with the national government for greater transparency about mining revenues and more strategic investment of these revenues in the region. It could mean becoming a true development partner with a region and setting a model for what sustainable mining might look like.

Shared social value is a new business narrative that takes us beyond FPIC and social license to a higher level of awareness within corporate culture of the responsibilities that should come with operating an ethical enterprise. At a time when public trust in institutions, and particularly multinational corporations, is at an all-time low, smart CEOs see shared social value as perhaps the only way companies may secure and sustain trust and be part of building a more humane and just form of capitalism. By embracing the shared value narrative, the company still seeks to play a major role in determining how value is defined. While conceding the legitimacy of the ethical question, the company retains control of the value narrative. But this is a pressure point where peacebuilders and activists can push for more substantial change.

Fiscal justice: transparency and taxation

For much of their history, extractive industries have enjoyed the advantage of relative obscurity. They operate in remote locations the public seldom sees. Their brands are unknown and their products sold business-to-business. It is only within the last twenty-five years that economists have blamed extractive industries for the "resource curse," environmentalists for egregious dumping of waste products, and human rights activists for gross human rights violations. As citizens are demanding much greater transparency and accountability from their leaders and institutions, mining, given its sizable economic footprint, has become a major target of scrutiny.

After a 2003 World Bank report called for an end to investments in coal and other major extractive projects, UK Prime Minister Tony Blair spearheaded the Extractive Industry Transparency Initiative (EITI). This initiative required governments and companies to subscribe publicly to its Ten Principles that included transparent corporate governance, human rights, sustainability, sound risk management, zero harm on worksites, and continuous environmental improvement in order to receive certification and become an EITI member.[2] Reviews of this initiative are at best mixed, with many countries embedding the principles in policy but failing to monitor implementation. Nonetheless, this initiative has offered advocates a forum in which to pressure both governments and companies on failed performance.

Building on the momentum of the EITI process, one of the most significant accomplishments of the last twenty years was the United States' passage of Section 1504 of the Wall Street Reform Act in 2010 (HR 4173, 111th Cong. (2010)). This tiny amendment was added to a massive financial reform law, despite strong opposition, especially from oil companies. It requires that all oil, gas, and mining companies publicly traded on the US Stock Exchange publicly report their financial transactions with any government in the world down to the project level. Since 95% of extractive industries globally are traded on the NY Stock Exchange, it has almost global coverage. Soon after, similar pieces of legislation were passed in the UK, the EU, Norway, and Canada. Companies soon realized that it was no longer possible to avoid these reporting requirements and quickly adjusted their public reporting to reflect the full scale of their financial dealings with countries around the world. Overnight, billions of dollars of royalties, profit, and miscellaneous financial transactions were made public. It was now possible for citizen groups across the world to actually see the level of revenue being received in both taxes and royalties from large international mining companies and to ask probing questions of their governments about how those funds were being allocated to national budgets, opening up more opportunity to recognize corruption. For companies that paid taxes to their host governments, the disclosures shifted attention to the governments' misappropriation of funds. Bolivia is a case in point. It renegotiated its contracts with many extractive firms and now enjoys massive government surpluses that it can dedicate to its own development.

A corollary to transparency is taxation. While transparency legislation in the United States and Europe is a considerable advance, it does not necessarily put an end to tax evasion and tax avoidance. Companies that engage in these practices generally under-value and under-report their levels of production and export value in order to decrease tax exposure in the exporting country. Then they take the profit from the sales of these commodities and move it through various offshore tax havens until it ends up in the one with the most favorable tax rates, thereby minimizing their tax exposure both in their country of origin and in the country where the company is headquartered. These practices bilk governments and citizens of billions of dollars that could be used for social programs, infrastructure, and productive investments in their home countries.

The release of the Panama Papers in 2016 provided a glimpse of the scale of these practices, but document only one firm in one tax haven (International Consortium of Investigative Journalists n.d.). It was the tip of the iceberg on tax evasion and avoidance. While governments have decried these practices and the OECD has sponsored a major initiative to address them, little has been done to shut down these tax havens. Important work is needed at the global level to confront this issue, change the way taxation is included in contract negotiations, reclaim the value of diverted funds, and reinvest them in development initiatives in their countries of origin.

Fiscal justice offers significant opportunities for building more accountable governance and promoting peace. Most peace accords involve the assurance of a more promising future for all parties. That requires financial investment. Funds lost to corruption, badly negotiated mining contracts, or tax evasion or avoidance represent an enormous missed opportunity. In many countries, the value of funds lost in this fashion dwarfs what is provided by donor governments in foreign aid or returned by companies in corporate social responsibility projects. Capturing this lost value and rechanneling it to development finance in poor communities where it originated could go a long way toward ensuring a more peaceful future.

Peacebuilding and extractive industries: time for new strategies and tactics

The triggers for mining-related conflict are complex, varied, and context specific, yet for the most part they have roots in individuals' and communities' sense that they have lost dignity and their ability to thrive economically and socially given the scale of disruption mining can unleash. Exercising their voice, they may first resist nonviolently, blocking a road or picketing a company office. If ignored, they may incite violence and wider conflict. Companies have historically looked to the state to intervene and neutralize the threat and defend their "right" to exploit, but increasingly today the use of coercive state power is seen within the industry as undermining the long-term viability of a project. They are now accountable to both shareholders and a new class of stakeholders who have invested in learning their business, imposing new norms, and demanding higher standards. In response, companies are seeking new ways to engage, gain legitimacy, and build trust.

One of the places that road has led is the Vatican. Over the last several years, the Vatican has convened some five meetings on mining issues with CEOs of some of the world's largest corporations. The church, through its global reach, has also heard and convened scholars, activists, peacebuilders, and other representatives from communities affected negatively by mining and enabled them to present criticisms, grievances, and general recommendations for improvements. Companies have enormous financial, political, and social power because of their sheer size. They have used this power historically to secure rights and legislation that allow them to advance their interests, often at odds with the interests of local communities. Levelling the playing field between companies and local communities is a tall order. The church is one of the few institutions in the world that has the power to challenge extractive industries on an equal footing, pose thorny moral and ethical questions, and propose bold reforms in policy, practice, and institutional culture. It has exercised this moral authority on behalf of aggrieved mining communities at local and national levels, often with positive outcomes. While it has begun to engage extractive industries at the global level, it is still finding its voice and role and has not yet proven to be as effective as it could be.

In today's world, with the ever-growing demand for metals, minerals, and energy, it is probably unrealistic to assume that governments are going to forego mining revenue as they build their national economies. Outright denial of mining rights is rare. In most countries, a strong case must be made and backed by considerable public pressure to cancel a major project. If we must, therefore, live with some forms of extraction, how might peacebuilders maximize the Catholic community's potential to tame it? Traditional grassroots peacebuilding and human rights strategies alone will not be enough. New strategies, tactics, and alliances will be required that build upon the gains in norms and policy achieved over the last three decades. The following are some recommendations.

Find allies with expertise

One of the reasons the church has fallen short of being effective has to do with its lack of sophistication about many of the business, technical, legal, and financial issues specific to mining. Many recommendations are in areas in which the industry has already moved ahead and made significant investments in improvement. Church representatives are often unaware of many of these advances and may risk appearing uninformed about the strategic debates currently underway within the sector. Another reason is that the church has concentrated more on confronting the industry at the mine site level rather than at the corporate or policy level. In reality, many of the problems facing communities can be far more effectively addressed through policy, regulation, and proper government oversight. The goal of the church's important grassroots work should be to link grievances and injustices at the community level to patterns of abuse at the national and global level and then turn the observation of these patterns into very specific policy recommendations to the industry at appropriate fora.

Oxfam's No Dirty Gold campaign has highlighted the importance of church leadership in Peru and El Salvador, and Publish What You Pay has pointed to the Catholic Church as a major partner in its efforts to increase transparency (van Oranje and Parham 2009, 36-37, 91-92). These examples show how dialogue and cooperation between civil society and the church can amplify the work of both for justice and peace in the mining sector.

Integrate peacebuilding into international development

Phil Vernon's (2015) report for International Alert provides a compelling analysis of how peacebuilding and economic development might benefit from a more integral relationship. Peacebuilders are generally seen by the development community as bystanders waiting for a conflict, rather than as partners on a common journey. Vernon's report claims that most conflict originates in some kind of economic challenge to the status quo that spins out of control. Therefore, effective development and peacebuilding require a sophisticated appreciation of local political economies, stakeholders, and their interests. Vernon argues that peace is best achieved when all parties to a conflict can find common ground and achieve an agreed-upon peace dividend. Peacebuilding that occurs in parallel or at odds with development actors is likely to fail or be perceived as marginal. The challenge for all advocates is to be clear about their complementarities in terms of knowledge, experience, competencies, and agendas. In the end, their shared interest is peace. Collaboration toward shared goals only empowers them both.

Embrace sustainability as a peace principle

Many mining companies have embraced sustainability as a principle driving future performance. They are attuned to the global focus on the Sustainable Development Goals; however, achieving real sustainability is an expensive challenge for mining companies. Mining requires major capital investments and access to abundant supplies of water and energy. Companies today recognize that in a world concerned with climate change and resource depletion, they must rethink their business models, invest in improved technology for environmental protection and cleanup, and assume greater responsibility for their impact on people and the land.

Environmental science within the industry has advanced dramatically over the last three decades. Peacebuilders would benefit from joining with environmentalists whose knowledge of these advances, and their shortcomings, could give them greater leverage with companies by providing the information needed to challenge company practices and governments' failures to hold companies accountable. Such protections for the environment could go a long way toward mitigating or preventing conflict between communities and companies as well as between communities who might otherwise experience rising tensions from resource scarcity. Mining will probably never achieve a perfect level of sustainability, but pressuring companies to adopt a shared-value approach and engage communities, governments, and stakeholders more transparently and proactively will push them to innovate and move in new directions. The Development Partner Initiative is an effort launched within the industry to foster precisely this kind of thinking, and peacebuilders should try to embrace and advance such efforts so that they become part of mainstream thinking.[3]

The logic of sustainability is central to Pope Francis's call to action in *Laudato Si'* which challenges individuals, governments, and businesses to work toward the creation of an economic order concerned with stewardship of the earth and dignity and integral human development for all humankind. Catholic actors can apply those ideas more effectively if they build off the present but imperfect industry-centered efforts toward sustainability. Doing so would also allow them to voice ideas like those from *Laudato Si'* that critique the technocratic paradigm that drives the mining industry and encourage changes and solutions that go beyond just technological innovation, such as supporting communities' right to say no to mining.

Broaden the free, prior, and informed consent framework

FPIC and transparency are perhaps the most powerful tools available to raise ethical questions and drive social change in the mining sector. The fact that ICMM has incorporated a free, prior, and informed consultation principle into their policy guidelines is progress, but insufficient progress. In much of the world, FPIC is still not a right recognized by governments or taken seriously by industry. Communities must be afforded the right to determine if the benefits of a mining project are worth sacrificing the investments their families have made in a region over generations. Peacebuilding groups should continue to press governments, investors, and civil society groups to not only support FPIC but enhance it. They should support and join efforts to legislate it at the national level and broaden the basis on which communities may assess risk by including not only environmental but other economic, livelihood, and human rights considerations. Communities should be given access to a due process review of any proposed project in an appropriate venue. FPIC is a normative framework that can rebalance relations in ways that may serve communities and avert unneeded division and conflict, but only if it is more deeply entrenched and more robustly developed and only if consultation becomes consent. The grassroots breadth of church actors combined with the church's vertical reach to higher levels of authority, as well as its legitimacy and moral authority, give Catholic peacebuilders particularly strong capacity for effecting these sorts of changes if they can marshal it effectively.

Push for improved transparency

The situation is similar with transparency. A source of conflict in many mining regions is under-investment by both the state and private sector in the broader development of the region, creating the perception that the promised benefits of mining are never realized. The transparency reforms that have forced companies to report profits, royalties, and all other payments at the project level are enabling tax authorities to audit accounts more carefully and generate more revenue. They have also enabled citizens to influence legislatures to redirect some portion of previously generated tax revenue back to their region of origin. The Peruvian government had for years been holding millions of dollars of such funds in regionally designated accounts in its Central Bank because they felt regional governments lacked the capacity to manage these large sums. Now, as part of a national decentralization process, it has begun releasing those funds to the regions to underwrite a variety of public services and infrastructure projects. And Burkina Faso has now legislated that one percent of annual revenue from gold mining be reinvested in mining communities. That may seem low, but this is a major source of tax revenue for a poor country, and it's a start. The precedent matters and allows regional governments to press for disbursements and challenge for increases in the future. This is a space where investments need to be made by governments and companies in building local governing capacity to support local management of such large sums.

Making the flow of this capital visible enables civil society actors to press for fiscal justice for communities through a proportionate share of revenue in equitable and sustainable development for their regions. But in many cases this information is not put to effective use. If contracts are not made more equitable, and taxes paid and allotted fairly, then any sins revealed by transparency initiatives pass without accountability. Leveraging that information to encourage investment planning in the early phase of project development may have the important indirect effect of minimizing future tension and conflict aggravated by economic hardships. While such fiscal policy work may seem tangential to traditional peacebuilding and human rights work, it is necessary to give tools like FPIC and transparency real weight.

Demand hard law when soft law fails

While some countries have binding laws on transparency and FPIC, for the most part companies are only governed by voluntary compliance frameworks.

Hence, company response has been uneven. For these mechanisms to be as effective as they can be for building peace and justice, they need to be written into legislation and actually enforced. As noted above with regard to FPIC, the vertical reach of the Catholic community gives it particular capacity for policy advocacy that connects community-level experiences and needs with regulatory and legislative bodies. This is not a domain in which many in the peacebuilding or broader activist community have been active. Yet it could be fruitful for such groups to focus some of their energies here in the future.

Conclusion

The mining sector is large and complex, as are the issues that surround it. The sector is changing, albeit under pressure. Leading companies are looking for advice and allies to help it change in the right ways. Engaging this momentum for change in a productive way is a unique opportunity for Catholic peacebuilders. What must be done to use it successfully?

To begin, peacebuilders and human rights advocates must begin to think outside their comfort zone and become educated about the sector, its history, culture, operations, financing, politics, and risks. Success in driving change will depend in large measure on the ability to be a challenging advocate with a strong moral narrative, an evidence-based critique, and technically and financially sophisticated propositions for reform. It will also require finding allies within the industry to begin and sustain dialogue, even as strong, credible, yet constructive critics.

The moral authority of the church represents a very significant brand and reputational risk to mining firms. Through its global presence, the Catholic Church is very often on the frontlines in mining conflicts. Industry leaders recognize the influence of the church on political leaders across the world, which explains why they have attended five meetings at the Vatican. Going forward, the church has to present a much more unified and effective front to industry. It must find ways to transform the multiple grievances it hears from its grassroots constituents into a coherent agenda of proposals to present to industry leaders. It must invest in building an internal structure that allows it to operate coherently on the local, national, and global levels. It must assemble a community of diverse advisors with deep understanding of the financial, technical, and operational aspects of the industry to assist in formulating its proposals and representing its views in formal sessions. It must develop an agenda of issues that are priorities for reform at the local, national, and industry levels and make common cause with the leading international NGOs and civil society organizations that have a long track record of commitment and achievement in this sector. It must give serious consideration to how it might use its powerful moral and ethical voice in support of the cutting-edge reform agendas currently on the table that hold the most promise of success. Its secular partners can provide excellent counsel on the opportunities, complexities, and politics surrounding each of these issues.

The good news is that the mining industry knows that it has to do better. Its past history of exploitation and abuse is a legacy liability in the twenty-first century. It must reinvent itself in order to survive operationally and financially. It is beginning to invest in research and development of new technologies. The more enlightened firms have embraced principles of sustainability, shared value, stakeholder consultation, and human rights and are trying to figure out how to include these ideas in their multi-year strategic and operational plans. Others are embracing a mindset shift toward becoming fully engaged development partners—moving from thinking narrowly inside the fence to thinking broadly outside the fence.

These are trends that, if harnessed, bode well for change that can benefit millions of people in countries in which mining is the dominant industry. The church and its field-based peacebuilders hold the potential to bring a moral framework to these industry reflections that could enrich the lives of millions, reduce corruption and conflict, and advance sustainability across the globe.

Notes

1 Majors include such companies as Anglo American, AngloGold Ashanti, Glencore, Newont, Rio Tinto, Valle, BHP, et al.
2 See https://eiti.org.
3 See http://www.dpimining.org.

References

Australian Associated Press. 2019. "BHP and Rio Tinto Join Push for Indigenous Voice to Parliament." *The Guardian*, January 30, 2019. https://www.theguardian.com/australia-news/2019/jan/31/bhp-and-rio-tinto-to-join-push-for-indigenous-voice-to-parliament.

International Consortium of Investigative Journalists. n.d. "The Panama Papers: Exposing the Rogue Offshore Finance Industry." Accessed March 22, 2021. https://www.icij.org/investigations/panama-papers.

International Council of Mining and Metals. n.d.a. "Indigenous People and Mining." Accessed March 22, 2021. http://www.icmm.com/en-gb/members/member-requirements/position-statements/indigenous-peoples.

International Council of Mining and Metals. n.d.b. "Mining Principles." Accessed March 22, 2021. https://www.icmm.com/en-gb/members/member-requirements/mining-principles.

Porter, Michael, and Mark R. Kramer. 2001. "Creating Shared Value." *Harvard Business Review*. https://hbr.org/2011/01/the-big-idea-creating-shared-value.

Russell, Grahame. 2019. "Es posible alcanzar la justicia en Canadá o en Guatemala por la represión vinculada a la minería de Hudbay Minerals/CGN?" *Rights Action*. July 23, 2019. https://mailchi.mp/rightsaction/es-posible-alcanzar-la-justicia-en-canad-o-en-guatemala-por-la-represin-vinculada-a-la-minera-de-hudbay-mineralscgn.

US Geological Survey. n.d. National Minerals Information Center. *Minerals Yearbook: Volume I —Metals and Minerals*. Accessed March 22, 2021. https://www.usgs.gov/centers/nmic/minerals-yearbook-metals-and-minerals.

van Oranje, Mabel, and Henry Parham. 2009. *Publishing What We Learned: An Assessment of the Publish What You Pay Coalition*. London: Publish What You Pay. https://eiti.org/files/documents/Publishing%20What%20We%20Learned.pdf.

Vernon, Phil. 2015. *Peace Through Prosperity: Integrating Peacebuiding Into Economic Development*. London: International Alert. https://www.international-alert.org/publications/peace-through-prosperity.

World Bank. 2003. *Striking a Better Balance: Volume 1. The World Bank Group and Extractive Industries*. Washington, DC: World Bank. https://openknowledge.worldbank.org/handle/10986/17705.

World Economic Forum and Global Agenda Council on the Future of Mining and Metals. 2015. "Voluntary Responsible Mining Initiatives: A Review." White Paper. *World Economic Forum and Resolve Solutions Network*. http://www3.weforum.org/docs/Voluntary_Responsible_Mining_Initiatives_2016.pdf.

16 Hardrock mining, climate change, and conflict: reflections through the lens of Catholic social thought

William N. Holden &
Caesar A. Montevecchio

Hardrock mining and climate change

Metals such as copper, gold, lead, nickel, silver, and zinc are products of "hardrock mining," which is often overlooked in discussions of climate change. As Benjamin Auciello (2019, 6) points out, analyses and policies that address extractive industries and climate change tend to focus on fossil fuels, rarely foregrounding metal and mineral mining. While hardrock mining may not extract products that emit greenhouse gases, it has important interfaces with climate change, in how it relates to climate vulnerability, and in how it is central to low-carbon growth. This chapter considers the relevance of these interfaces for conflict and peace and analyzes them through the lens of Catholic social thought and praxis. This includes climate injustice and disparate climate impacts between the Global North and Global South; "extractivism" and harmful senses of development; violence against human rights and environmental defenders; and unique questions of uranium mining and the specter of nuclear arms.

Metals mining and climate vulnerability

Mining is an activity with substantial environmental consequences, such as siltation, contamination, and depletion of water, air pollution, and deforestation (see Holden and Jacobson 2012, 59–76). Most are exacerbated by climate change. One of the most serious is acid mine drainage. Rocks containing desirable minerals are often found in sulfide ore deposits; when these are exposed to water and oxygen, they release sulfuric acid. As this continues, heavy metals such as arsenic, lead, and mercury can be mobilized.

The principal method of managing acid mine drainage is to place mining wastes (or "tailings") under water behind a tailings dam in an anaerobic environment. For this to be effective water must always be over the material; should a tailings dam fail, contaminants would be released. Since the geochemical processes of acid mine drainage operate on a geologic time

DOI: 10.4324/9781003094272-16

scale, these dams require perpetual attention. And over time, climate change will present more and more problems, such as by bringing more heat waves, droughts, and wildfires.[1] There will also be stronger tropical cyclones and heavier rainfall increasing the risk of dam failures (National Academies of Sciences, Engineering, and Medicine, 2016).

Mining companies will be increasingly frustrated in searching for technological solutions to these and other environmental risks from climate change. In planning mines and waste storage, mining engineers will find it more and more difficult to rely upon stationarity, the idea that precipitation levels can be measured within upper and lower limits determined by climatic history. With climate change, stationarity has become unreliable as a principle of water management (Milly et al. 2008). Mine waste storage facilities must now be constructed to accommodate worst-case scenarios, not what the climatic records indicate as the highest rainfall levels likely to occur; given the uncertainty surrounding the rapidity of climate change, this determination will be difficult (Holden 2015, 11).

Metals mining and low carbon growth

The prevention of catastrophic climate change requires a transition away from fossil fuels and towards renewable energy. But dozens of mined metals are employed in renewable technologies (Dominish, Florin, and Teske 2019, 16). For example, aluminum and copper are required for wind turbines, photovoltaic panels, and rechargeable car batteries; silver, selenium, and tellurium for photovoltaic panels; and cobalt, lithium, and nickel for rechargeable car batteries. Auciello (2019, 13) points out: "Mining companies stand to benefit by positioning themselves as key actors in the transition. They are taking advantage of the projected increase in critical metals to greenwash and justify new operations, regardless of whether their output will be used in renewable energy technologies or not."

Another mineral which might see increased demand as the world makes a transition away from fossil fuels is uranium, a fuel for nuclear power plants. The advantage of nuclear energy lies in the tremendous amount of energy generated by a relatively small quantity of uranium. Uranium has a high energy density per unit of fuel and one *gram* can produce as much energy as three *tons* of coal (Percebois 2003, 102). Ecologist James Lovelock has declared nuclear energy the only immediately available energy source not causing global warming (Marshall 2005, 1169). A 676-megawatt Canadian Deuterium Uranium reactor will reduce carbon dioxide (CO_2) emissions by about 1.2 million tons per year if it displaces a coal powered plant (Morrison 2001, 46).

Since France produces approximately 80% of its electricity from nuclear power, it has the lowest CO_2 per unit of gross domestic product of any country in the world. If the world's CO_2 emissions were as low as France's, they would be reduced by half (Richter 2006, 15). Indeed, the construction of 3,000 to 3,200 new nuclear power plants worldwide by 2050 could stabilize CO_2 concentrations at twice pre-industrial levels, notwithstanding a 50% increase in world population and a 50% increase in per capita energy consumption (Macfarlane 2003). When Germany retired all its nuclear power plants after the 2011 accident at Fukushima, it saw a significant and rapid spike in CO_2 emissions. By displacing fossil fuels, nuclear power has potential to slow down, or possibly even avert, catastrophic climate change.

There are risks in nuclear power, as demonstrated by the accidents at Fukushima and at Chernobyl in 1986. At Chernobyl, the official death toll was 47, although some estimates go as high as 4,000 when considering radiation effects. At Fukushima not one person died and, in the words of David Wallace-Wells (2019, 183), "Had none of the 100,000 living in the evacuation zone ever left, perhaps a few hundred might have ultimately died of cancers related to the radiation." But for comparison's sake, it is estimated that 10,000 people die every day from the particulate matter emitted by burning carbon (183). At the very least, this comparison calls into question whether the risk of such events is a worthwhile tradeoff for lowered carbon emission.

Unfortunately, one of the largest disadvantages of nuclear energy is the environmental impact of uranium mining. This is where nuclear energy can have some of its most serious environmental effects. Uranium mining causes the same environmental problems as other types of hardrock mining, but also involves radioactivity. Uranium mining can release radionuclides into the atmosphere through the crushing and grinding of ore, from fugitive dust emissions, and from the release of radon gases (Thomas and Gates 1999, 527). Moreover, uranium radioactivity and acid mine drainage are not problems that exist *simultaneously* but *independently* of each other; rather, they have a perverse *synergy* in that acid mine drainage leads to the further dissolution of radionuclides, enhancing the radioactivity given off by the uranium (Fernandes et al. 1998; Ripley et al. 1996, 201). According to Saleem Ali (2003, 109), "In terms of both short-term and long-term environmental impact, uranium mining is by far the most environmentally problematic of any mining activity—owing to the simple fact that radioactivity of the ore presents an intangible that cannot be chemically mitigated."

Hardrock mining, climate change, and challenges to peace

Mining is a leading cause of socio-environmental conflict (Auciello 2019, 17). As of August 21, 2019, the Environmental Justice Atlas documented 2,865 cases of socio-environmental conflict, with 587 related to mining; of these, 260 are related to the extraction or processing of metals. That is a lower number than combined fossil fuel-related conflicts (oil, 279; coal, 212; gas, 153), but metals-mining conflicts could well overtake fossil fuel-related conflicts if the extractive frontier expands towards the metals necessary for renewable energy (Environmental Justice Atlas n.d.). And though it does not register as socio-environmental conflict, the risk of nuclear arms that is presented by uranium mining is another important challenge for peace. The following considers some of the specific factors involved in these situations and responds to them with insights from Catholic social thought and praxis.

Pope Francis is the first pope from the Global South, and his affection for the environment became apparent when *Laudato Si'* was released. Its predominant theme is "integral ecology" and how this responds to the dangers and injustices created by environmental degradation, including the seriousness of climate change. Francis (2015, §26) affirms the pursuit of renewable energy: "There is an urgent need to develop policies so that, in the next few years, the emission of carbon dioxide and other highly polluting gases can be drastically reduced, for example, substituting for fossil fuels and developing sources of renewable energy." Although Francis does not address using nuclear energy as a method of providing low-carbon electricity, he does intimate that the use of less harmful alternatives or other short-term solutions, which could include nuclear energy, may be necessary (§165). But throughout, Francis's teaching on the environment demonstrates constant awareness of the possible injustice created as those in the Global South bear the brunt of the impact of worsening climate change and of the shortcomings of the technocratic solution models attached to mining and energy.

North-south climate injustice and the need for subsidiarity and solidarity

Relying on hardrock minerals to alleviate climate change's effects is highly problematic as it perpetuates an economic and cultural paradigm inseparable from injustice, conflict, and environmental degradation. It is also a paradigm that maintains colonial disparities between the Global North and the Global South, as much of the mining of those minerals occurs in the latter while the consumption of them will occur in the former. This can be approached through the perspective of climate injustice, how some people enjoy the benefits of energy use and other emissions-generating activities, while others disproportionately suffer the burdens of climate change (Bell 2013, 190). Specifically, this means that locations in the Global South where the minerals necessary for a transition to a low-carbon future are found bear an unjust and unequal amount of their cost (Auciello 2019, 7).

The Democratic Republic of the Congo (DRC) exemplifies these dynamics. As outlined in the chapter by Rigobert Minani, privatization of the mining industry in the DRC amplified economic and political insecurity and significantly exacerbated conflict. These problems were particularly concentrated in the eastern part of the country, where most of the country's cobalt reserves are located and where mining companies engaged with and enriched rebel groups that had gained more influence in the region than the national government for a period beginning in the late 1990s. Cobalt is essential for renewable energy, and as demand for it has rapidly increased, tremendous human and environmental costs have been imposed on the people of the DRC (Noriyoshi 2020). This mining supports solutions for climate change in the Global North, but does so while passing a great burden to the Global South, disempowering already vulnerable populations and worsening violent conflict.

Pope Francis's teaching has shown great concern for this kind of disparity. *Querida Amazonia,* Francis's apostolic exhortation following the 2019 Synod on the Amazon, applies *Laudato Si*'s ecological framework to the environmental justice challenges of the region. The document predominantly focuses on indigenous peoples—both their suffering and victimization, and the inherent value of their heritage, which can be a source of wisdom for those in the throes of globalized capitalism and neoliberal ideas of development. And it does so while placing that plight under the rubric of colonialism, as well as connecting it to global environmental problems (Francis 2020, §14, 48).

Francis observes that the "techno-economic" power structures rooted in the Global North are the primary drivers of extractive and other projects endangering the Amazon and are a threat to "overwhelm not only our politics, but also freedom and justice" (§52). Part of his solution, rooted in the principle of subsidiarity, is to empower national leaders to protect ecologically sensitive areas against "huge global economic interests" that would otherwise "internationalize" the Amazon (§50). He supports free, prior, and informed consent, the right of communities to say no to projects and propose development alternatives, and the need for legal frameworks to protect and enforce such rights (§51–52). But more deeply, he appeals to the wisdom of indigenous cultural heritage and local systems of knowledge as sources for helping to make better decisions about how land is developed and used. Such sources can help awaken a sense of ecological spirituality, engendering greater ecological care (§42).

In his chapter, Vincent Miller outlined a way for the Catholic Church to use its ecclesial networks to help develop solidarity between the Global North and supply chain origin points in the Global South. If *Querida Amazonia's* "ecclesial dream" of a church that learns from indigenous peoples in the Amazon can be actualized (§28–40), then the networks of solidarity described by Miller could also serve as channels for developing greater care for ecologically vital but threatened areas like the rainforests of the Amazon and the Congo Basin. Forming greater solidarity in this way will not solve climate justice, but it can be one humble step along the way. It could create a greater sense of responsibility among those in the Global North for environmental damage impacting everyone, but impacting the Global South faster and more acutely.

Extractivism and integral human development

As a method of forestalling climate change, focusing on the extraction of hardrock minerals in contradistinction to continued extraction of fossil fuels exemplifies the dominance of a paradigm referred to as extractivism. Naomi Klein (2014, 169) describes it as:

> ...a nonreciprocal, dominance-based relationship with the earth, one purely of taking. It is the opposite of stewardship, which involves taking but also taking care that regeneration and future life continue. Extractivism is the mentality of the mountaintop remover and the old-growth clear-cutter. It is the reduction of life into objects for the use of others, giving them no integrity or value of their own—turning living complex ecosystems into 'natural resources.'

Extractivism is based on continuously removing more, and more, and more from the earth. As climate change poses an existential threat to humanity, we make the transition from extracting fossil fuels to extracting hardrock minerals to advance renewable energy technologies, but we never stop extracting. The places where the minerals are found become what Auciello (2019, 7) calls "new zones of sacrifice" while "the extractivist model expands, now masked behind the veil of meeting the material needs of the transition." A substantial impetus to extractivism has been the ascendency of neoliberalism, an aggressive, no-holds-barred, version of capitalism described by Gerardo Rénique (2006, 37), writing in a Latin American context, as *capitalismo salvaje* (savage capitalism). It has been instrumental in empowering the multinational corporations running major mining operations and in enabling the system allowing mined goods to flow through the global supply chain and acquire separation from the harms done at local mining sites (see Princen 2002).

The World Bank has encouraged the adoption of neoliberal policies and it has also been an impetus to the promotion of mining in the Global South (Holden and Jacobson 2012, 41). Two influential World Bank publications in the early 1990s made this pattern apparent. *The Strategy for African Mining* (1992) argued against public ownership of mining companies in developing countries and emphasized mining projects aimed primarily at production for export. Two years later, the Principal Mining Officer of the World Bank published "Strategies to Attract New Investment for African Mining," Strongman 1994, which made it clear that the World Bank's emphasis was on opening up the mining sector in Africa to foreign direct investment by multinational mining companies (Strongman 1994). Although these reports were focused on Africa, they had an impact on mining codes across the entire Global South. For example, the Philippine Mining Act of 1995 contained incentives to encourage mining such as a four-year income tax holiday, tax and duty-free capital equipment imports, value-added tax exemptions, income tax deductions for losses, and accelerated depreciation (Holden and Jacobson 2012, 41). It also guaranteed the right of repatriation of the entire profits of the investment and guaranteed no mining project would ever be expropriated. Mining codes like this justify these incentives based on the premise that economic activity generated will, somehow, trickle down to the poor. This is a highly problematic strategy. Their tax incentives produce minimal revenue for their host country (188), modern capital-intensive methods of mining produce minimal employment (187), and the environmental effects of mining degrade natural resources relied upon by the rural poor and thrust them from subsistence into destitution (205).

What the extractivist paradigm never contemplates is a cessation of extraction. The idea of removing *less* from the earth and getting *more* out of what has already been extracted—or changing consumption patterns—is never considered. It is in this context that recycling becomes crucial. "The billions of tons of material already mined and circulating in cities and factories or lying in landfills," wrote Payal Sampat (2003, 111), "can serve the same functions as underground ore, with far fewer ecological costs." Many of the metals required for renewable energy are highly recyclable (Dominish, Florin, and Teske 2019, 16). Instead of mining virgin ore deposits, with all the environmental effects involved, these metals could be acquired through recycling, which is also more labor-intensive than mining and thus creates more jobs (Auciello 2019, 22). Uranium can also be recycled. When spent nuclear fuel is removed from a nuclear reactor it still has some fissionable material left and can be subjected to a process separating out reusable uranium and plutonium from non-recoverable waste products. Fuel recycling reduces the volume of final waste by a factor of five, but the cost of a kilogram of reprocessed fuel is approximately nineteen times more than the cost of a kilogram of natural uranium (Percebois 2003, 102). Other types of recycling are similarly expensive. None of these outcomes aligns with neoliberal extractivism or the consumption habits of those in the Global North.

The argument about whether to use fossil fuels *or* mining-dependent renewable energy technologies may be missing a bigger point: on a planet with finite natural resources, is adherence to an economic paradigm requiring continuous growth acceptable? The question is especially urgent given how the extractivist model driving the production of these resources also drives continued poverty, environmental vulnerability, and violent conflict in the Global South.

To Pope Francis (2015, §109), the neoliberal model is problematic because "by itself, the market cannot guarantee integral human development and social inclusion." Modern neoliberal capitalism is "the mindset of those who say: let us allow the invisible forces of the market to regulate the economy and consider their impact on society and nature as collateral damage" (§123). This is unacceptable because "the environment is one of those goods that cannot be adequately safeguarded or promoted by market forces" (§190). Francis's ideas echo those found in a letter from the Latin American Provincials of the Society of Jesus (1997, 48) articulating their opposition to this economic paradigm: "Neoliberalism is a radical conception of capitalism that tends to absolutize the market and transform it into the means, the method, and the end of all intelligent and rational behavior."

A hallmark of contemporary Catholic social teaching is that technocratic and market approaches cannot meet requirements for human flourishing. These approaches do not adequately factor in either the value of the environment as a good in itself or the conflicts associated with some economic practices. Given that conflict is related to mining in the Global South, and how conflict is compounded by effects of climate change, integral human development in the mining sector is closely related to peacebuilding. A pastoral letter from the Latin American Episcopal Council CELAM (Consejo Episcopal Latinoamericano) 2018 ties these points together well. The letter criticizes the extractivist model driving mining development for its "utilitarian mindset... in service to the idolatry of capital," and outlines this model's impacts on the environment, continued poverty, and socio-environmental conflicts (CELAM 2018, §28–46). As a remedy, the letter advocates for a "radical change in the understanding and practice of the economy" and a shift to new models of development in line with integral human development (§103–8).

Clemens Sedmak's essay in this book offers further analysis of mining and integral human development through the lens of *Laudato Si'*. According to Sedmak, the most important contribution that the ethics of integral human development could make in terms of the mining industry is a rejection of maximization of profits as an ultimate goal. He builds this claim from *Laudato Si*'s observation that profit is illegitimate when it does not advance the integral development of the human person or observe the preferential option for the poor, and that care for creation is inseparable from both of those goals. Integral human development, according to Sedmak, introduces a mandate for restraint that is absent from the neoliberal model of unlimited growth and which demands that the human flourishing of the communities impacted by mining be given priority over profit maximization. This would include ecological protection ensuring communities' lands are not unjustly despoiled. This would also curb global climate change and help prevent further environmental degradation.

Picking up the theme of Catholic social teaching demanding multifaceted, integral approaches to problems, such as with the model of integral human development, Gerard Powers (2017, 8) suggests the need to form a model of integral peacebuilding. Climate change drives mining; mining impacts conflict vulnerability, damages fragile ecologies on which communities depend, and introduces potentially harmful economic development strategies; and these instabilities make those communities more vulnerable to climate change effects, which in turn exacerbates conflict and creates potentially new sources for it. This chain indeed demands that peacebuilders practice integral approaches challenging the extractivist paradigm, and that are sensitive to climate change and married to development efforts in line with integral human development.

Violence against environmental defenders

In 2007, this chapter's first author was preparing to interview Bishop Gabriel Penate, Bishop of the Apostolic Vicariate of Izabal, in Puerto Barrios, Guatemala, about the local activities of a Canadian mining company. As the interview was starting, the Canadian ambassador to Guatemala telephoned Bishop Penate and informed him there were Canadian anti-mining activists in Guatemala and that he should not talk to them (Holden and Jacobson 2012, 50). This incident exemplifies how those with interests in mining operations are willing to employ intimidation and pressure against anyone resisting their agenda.

Such intimidation frequently transitions to direct violence. One of the most disturbing aspects of the socio-environmental conflicts associated with hardrock mining are assassinations and other forms of violence against human rights and environmental activists. In 2019, according to Global Witness (2020, 6), 50 of 212 murdered land and environmental defenders were opposing mining, the deadliest sector. Examples of this sort of violence include a 2018 incident in Malawi where thirteen people were killed by the police while protesting air and water pollution from a copper smelting plant (Global Witness 2019, 34). And in Colombia between 2015 and 2019, 44% of attacks against activists and land defenders were connected to five extractive companies (Business and Human Rights Resource Center 2020, 1).

According to Victoria Tauli-Corpus, the United Nations Special Rapporteur on the rights of indigenous people, killings of environmental activists worsen the problem of climate change:

> This violence is a human rights crisis, but it is also a threat to everyone who depends on a stable climate. Land and environmental defenders are among the best stewards of the world's great forests and biodiversity, and when their rights are trampled, it is often to make way for environmentally destructive logging, mining, or plantations. Protecting indigenous land rights defenders is therefore not only a human rights imperative—but also urgent to mitigating the climate crisis. (quoted in Global Witness 2019, 26)

Querida Amazonia (Francis 2020, §14) is direct on this issue, listing criminalization of protest and killings of people who oppose projects among the crimes committed by businesses in the Amazon. And it is more common to see the issue taken up by local Catholic bodies and on-the-ground actors. José Bayardo Chata Pacoricona's chapter describes how a significant part of Derechos Humanos y Medio Ambiente–Puno's legal defense work involves laws effectively criminalizing human rights and environmental protesters, and justifying violence against them. The Catholic Church in Colombia has lobbied the government to better defend social leaders and environmental protestors facing violence, threats, and displacement from mining operations (Fraser 2020). The Pan-Amazonian Ecclesial Network (2019) issued a statement urging national leaders and the international community to protect and defend peasants, indigenous peoples, and others facing violence and threats while trying to defend the earth and their communities. And in May 2020, CINEP/PPP, a Jesuit-run organization in Bogotá, held a virtual event to call attention to the way the COVID-19 pandemic was increasing the vulnerability of social leaders and bringing about an increase in violence against them.[2]

These examples show how regional, national, and local Catholic groups actively denounce and act against violence against those trying to defend their communities and ecologies against damaging activities like mining. They do so as an act of peacebuilding, and their praxis and witness provide vital contributions to the corpus of Catholic social tradition and join the global work of resisting climate change in real and substantive ways.

Uranium mining, nuclear arms, and global insecurity

As noted above, uranium mining may have the highest environmental cost of any form of mining because radioactivity cannot be mitigated as can some other mining byproducts. Though radioactivity is only one factor among many, the slow violence done to indigenous peoples in the American West by uranium mining testifies to this (see Fegadel 2020). But another issue is the specter of nuclear arms. William Perry, former U.S. Secretary of Defense, contends that the threat of nuclear war is now the greatest it has been since the Cold War (Haberman 2018). While most uranium is used for peaceful purposes, and despite major producers like Australia and Canada having strict measures to control uranium they sell, the risk of materials being channeled to weapons development is ever-present. Security of nuclear power facilities is a key concern. Reliance on nuclear energy to forestall climate change could increase the number of nuclear plants in countries without strong regulatory or security infrastructures to control weapons-grade materials at those plants. Vulnerability to cyberattacks, sabotage, or theft also remains significant in several nations already having nuclear capability (NTI 2020, n.p.).

Unlike many materials needed for clean energy technology, uranium is not predominantly mined in the Global South. Still, some of these countries, like Gabon or Niger, have been caught in the fray of the nuclear arms race. For example, after years under colonial rule and feeding into the nuclear ambitions of colonial powers, especially France, independent Niger asserted its independence by dealing with business partners such as Muammar Gaddafi, who became a broker to others (George 2019, 198). Furthermore, low market prices for uranium have prevented Niger from realizing significant economic benefits, leaving the country with a legacy of instability, fueling ongoing internal conflict, such as revolts by Tuareg people in northern Niger (199–201).

The Catholic record on nuclear weapons is well-established. Since *Pacem in Terris* (John XXIII 1963) the Catholic Church has been a leading voice on the morality of nuclear arms. The specifics of the church's teaching have evolved over the decades, but disarmament has always remained the ultimate goal (see Powers 2015). Pope Francis (2019) took the unprecedented step of calling even the possession of nuclear weapons immoral. While the church's teaching on disarmament is clear, uranium mining as an alternative to fossil fuels is more complicated. And Catholic voices have not always been strong about connecting local concerns with uranium mining, like effects of radioactivity on miners, with broader peace and security concerns like nuclear proliferation.

One recent analysis by a group of Catholic ethicists does demonstrate an integrated perspective about climate change, nuclear energy, uranium mining, and weapons proliferation. It concluded that nuclear power is unsuitable as a primary strategy for limiting climate change because of the time it would take to expand nuclear energy capacity and the diversion of resources from developing renewable energy (Biviano et al. 2016, 21–22). The same analysis also raised cautions about the byproducts of nuclear energy, specifically health effects from radioactivity, safety concerns about reactor malfunctions, and potential arms proliferation (19–20). This position helps establish a baseline of skepticism in Catholic thinking about nuclear energy.

And that skepticism can be seen in practice, if not necessarily with the full range of integrated perspective. For example, as noted above, the Navajo people of the southwestern United States have suffered a decades-long legacy of environmental, cultural, and social violence in connection to uranium mining. The US Catholic community in New Mexico, the center of many of these issues, has been active in resisting ongoing uranium mining and seeking justice for past harms. Over twenty years ago, the state's Catholic bishops issued a pastoral statement on global climate change, energy, and the unique related challenges for their state (Roman Catholic Bishops of New Mexico 1998). They argued that the climate crisis should not be compounded by mining projects worsening local ecological vulnerabilities or disproportionately affecting marginalized peoples. In 2015, the Office of Peace, Justice, and Creation Stewardship of the Diocese of Gallup offered testimony at a public hearing about a permit to resume uranium mining at Mount Taylor. The testimony was offered on behalf of "concerned people of faith in Catholic communities in Cibola County... who continue to suffer adverse health and environmental impacts from past decades of uranium mining, yet to be satisfactorily remediated to bring an end to continuing contamination of life-sustaining water, land and air in the Mount Taylor region" (Cecchini 2015). In both cases there is a clear focus on local impacts, and from the bishops, concern about the broader issue of climate change; however, neither statement took the step of also connecting those concerns to ethical questions about nuclear proliferation.

Catholic leaders also intervened on uranium mining in Malawi beginning in 2007. Paladin Energy, an Australian company, was in the process of opening the Kayelekera Uranium Mine for production, when the Catholic Commission for Justice and Peace of the Episcopal Conference of Malawi (CCJP) joined a group of NGOs filing an injunction alleging Paladin's environmental impact assessment failed to account for some of the unique hazards of uranium mining (CathNews.com 2007). The mine opened in 2009 and went into a care and maintenance phase in 2014.[3] In the year before it closed, Norwegian Church Aid and Catholic Commission for Justice and Peace Malawi (2013, 10, 28–30, 31–32) published a joint report focused particularly on economic issues with the Kayelekera project, and at several points it brought up the problem of displaced persons, repeatedly highlighting inadequate resettlement processes, losses of livelihoods, and community disruption as some of the chief problems of the country's mining sector. In a separate report, Catholic Commission for Justice and Peace Malawi (2014) compared mining displacement in Malawi with other countries in southern Africa. Like the 2013 report, this report developed legislation and policy recommendations, but did so specifically with regard to internal displacement caused by mining and expanded to consider environmental as well as social and economic issues. In recent years, such factors have substantially increased the potential for violent conflict in Malawi (Human Rights Watch 2016; Kishindo and Mvula 2017). Yet as in New Mexico, while Catholic leaders in Malawi dealt with important local and national issues, they did not bridge those issues with the problem of nuclear arms uniquely connected to uranium mining.

These two cases demonstrate how uranium mining, like nearly all hardrock mining, has had great difficulty meeting the standards of sustainability and justice required by Catholic social teaching. But they also show inconsistency in Catholic voices about connecting those critiques to the extra worry of nuclear weapons that gives questions about uranium mining distinct urgency. Nevertheless, even if it has not been consistently expressed, the framework of Catholic social tradition does offer a structure for integrating peace, sustainability, and justice in the analysis of uranium mining.

Conclusion

The hydrometeorological hazards posed by climate change pose a substantial threat to hardrock mines, and concerns about climate change may lead to an increase in metals mining because it is critical for clean energy. These realties have important implications for climate change, justice, and peace: disparate impacts on the Global South compared to the Global North; the maintenance of a neoliberal extractivist paradigm perpetuating injustices and yielding inadequate models of development; rises in violence against activists defending human rights and the environment; and questions about uranium mining, which introduces unique environmental hazards and risks contributing to nuclear proliferation.

Catholic social thought and the theory and praxis of peacebuilding offer responses to these issues. They speak loudly and clearly to the need for climate justice and greater solidarity with vulnerable peoples suffering violent conflict exacerbated by mining. Catholic social thought and peacebuilding also stringently interrogate the extractivist model and offer an alternate vision of integral human development. Catholic leaders and organizations actively engage the problem of violence against human rights and environmental defenders and take action to alleviate it. And lastly, the Catholic Church is a strong advocate for nuclear non-proliferation and disarmament. It is also a source of vigorous action at local and regional levels defending communities from the effects of uranium mining, though Catholic actors could do more to connect that advocacy and action to issues of nuclear weapons. Hardrock mining is very much at the center of climate change and conflict discussions, and Catholic teaching and peacebuilding can make important contributions to the conversation.

Notes

1 A recent example is the Carr Wildfire in California in July 2018, which nearly destroyed a water treatment facility built to mitigate acid drainage from the Iron Mountain Mine (closed in 1963).

2 https://www.cinep.org.co/Home2/component/k2/789-lideres-sociales-blanco-de-violencia-en-cuarentena.html.

3 When the mine was closed in 2014, the retrenchment of workers occurred with repressive force from armed security and left the mine without capacity to begin rehabilitation, monitor and mitigate radiation, or deal with a near-overflowing tailings dam (see Nyirenda, 2014).

References

Ali, Saleem H. 2003. *Mining, the Environment, and Indigenous Development Conflicts*. Tucson, AZ: University of Arizona Press.

Auciello, Benjamin Hitchcock. 2019. *A Just(ice) Transition is a Post-Extractive Transition*. London: War on Want and London Mining Network. https://waronwant.org/sites/default/files/Post-Extractivist_Transition_WEB_0.pdf.

Bell, Derek. 2013. "How Should We Think About Climate Justice?" *Environmental Ethics* 35, no. 2: 189–208. 10.5840/enviroethics201335217.

Biviano, Erin Lothes, David Cloutier, Elaine Padilla, Christiana Z. Peppard, and Jame Schaeffer. 2016. "Catholic Moral Traditions and Energy Ethics for the

Twenty-First Century." *Journal of Moral Theology* 5, no. 1: 1–36. https://jmt.scholasticahq.com/article/11306-catholic-moral-traditions-and-energy-ethics-for-the-twenty-first-century.

Business and Human Rights Resource Center. 2020. "Business and Human Rights Defenders in Colombia." https://media.business-humanrights.org/media/documents/files/Business__Human_Rights_Defenders_in_Colombia.pdf.

CathNews.com. 2007. "Malawi Catholics Seek Injunction Against Aussie Uranium Mine." April 9, 2007. https://cathnews.com/cathnews/12687-malawi-catholics-seek-injunction-against-aussie-uranium-mine.

Catholic Commission for Justice and Peace Malawi. 2014. *Land Displacement, Involuntary Resettlement and Compensation Practice in the Mining Sector: A Comparative Analysis of Legal and Policy Frameworks in Southern Africa.* https://www.ecmmw.org/new/wp-content/uploads/2014/12/Report-on-Land-Displacement-Inviluntarry-Resttlement-and-Compensation-Practices.pdf.

Cecchini, Rose Marie. 2015. "Testimony on Rio Grande Resources Standby Permit from 2010" (*Post-Hearing Public Comments*, Cibola County, New Mexico, May 2015). http://www.emnrd.state.nm.us/mmD/MARP/documents/MtTaylor_Rev10-1_PostHearingCommentsfromthePublic.pdf.

CELAM (Consejo Episcopal Latinoamericano). 2018. *Discípulos Misioneros: Custodios de Nuestro Hogar Común: Discernimiento a la Luz de la Encíclica* Laudato Si'. Bogotá: Consejo Episcopal Latinoamericano, CELAM.

Dominish, Elsa, Nick Florin, and Sven Teske. 2019. "Responsible Minerals Sourcing for Renewable Energy." Report prepared for Earthworks by the Institute for Sustainable Futures, University of Technology Sydney. https://www.earthworks.org/cms/assets/uploads/2019/04/MCEC_UTS_Report_lowres-1.pdf.

Environmental Justice Atlas. n.d. Accessed December 4, 2019. https://ejatlas.org/.

Fegadel, Averi R. 2020. *Toxic Colonialism and Green Victimization of Native Americans: An Examination of the Genocidal Impacts of Uranium Mining.* Ph.D. diss., University of South Florida. https://scholarcommons.usf.edu/etd/8194/.

Fernandes, Horst M., Mariza R. Franklin, and Lene H. Veiga. 1998. "Acid Rock Drainage and Radiological Environmental Impacts: A Study Case of the Uranium Mining and Milling Facilities at Pocos de Caldas." *Waste Management* 18, no. 3: 169–181. 10.1016/S0956-053X(98)00019-1.

Francis. 2015. *Laudato Si'.* http://www.vatican.va/content/francesco/en/encyclicals/documents/papa-francesco_20150524_enciclica-laudato-si.html.

Francis. 2019. "Address at the Hiroshima Peace Memorial." November 24, 2019. http://www.vatican.va/content/francesco/en/messages/pont-messages/2019/documents/papa-francesco_20191124_messaggio-incontropace-hiroshima.html.

Francis. 2020. *Querida Amazonia.* http://www.vatican.va/content/francesco/en/apost_exhortations/documents/papa-francesco_esortazione-ap_20200202_querida-amazonia.html.

Fraser, Barbara. 2018. "Church Urges Colombian Government to Protect Local Leaders Under Threat." Catholic News Service, September 5, 2018. Accessed in *National Catholic Reporter.* https://www.ncronline.org/news/world/church-urges-colombian-government-protect-local-leaders-under-threat.

George, William P. 2019. *Mining Morality: Prospecting for Ethics in a Wounded World.* Lanham, MD: Lexington Books, 2019.

Global Witness. 2019. *Enemies of the State? How Governments and Business Silence Land and Environmental Defenders*. London: Global Witness. https://www.globalwitness.org/en/campaigns/environmental-activists/enemies-state/.

Global Witness. 2020. *Defending Tomorrow: The Climate Crisis and Threats Against Land and Environmental Defenders*. London: Global Witness. https://www.globalwitness.org/en/campaigns/environmental-activists/defending-tomorrow/.

Haberman, Clyde. 2018. "'This Is Not a Drill': The Threat of Nuclear Annihilation." *New York Times*, May 13, 2018. https://www.nytimes.com/2018/05/13/us/nuclear-threat-retro-report.html.

Holden, William N. 2015. "Mining Amid Typhoons: Large-Scale Mining and Typhoon Vulnerability in the Philippines." *The Extractive Industries and Society* 2: 445–461. 10.1016/j.exis.2015.04.009.

Holden, William N., and R. Daniel Jacobson. 2012. *Mining and Natural Hazard Vulnerability in the Philippines: Digging to Development or Digging to Disaster?* London: Anthem Press. 10.7135/UPO9781843313960.

Human Rights Watch. 2016. "'They Destroyed Everything': Mining and Human Rights in Malawi." September 27, 2016. https://www.hrw.org/report/2016/09/27/they-destroyed-everything/mining-and-human-rights-malawi.

John XXIII. 1963. *Pacem in Terris*. https://www.vatican.va/content/john-xxiii/en/encyclicals/documents/hf_j-xxiii_enc_11041963_pacem.html.

Kishindo, Paul, and Peter Mvula. 2017. "Malawi's Land Problem and Potential for Rural Conflict." *Journal of Contemporary African Studies* 35, no. 3: 370–382. 10.1080/02589001.2017.1342784.

Klein, Naomi. 2014. *This Changes Everything: Capitalism vs The Climate*. Toronto: Knopf.

Latin American Provincials of the Society of Jesus. 1997. "Letter and Study Document on Neoliberalism in Latin America." *Promotio Justitae* 67: 43–60. https://www.sjesjesuits.global/media/2021/02/PJ_067_ENG.pdf.

Macfarlane, Allison. 2003. "Is Nuclear Energy the Answer?" In *Fueling the Future: How the Battle Over Energy is Changing Everything*, edited by Andrew Heintzman and Evan Solomon, 127–151. Toronto: Anansi Press, 2003.

Marshall, Eliot. "Is the Friendly Atom Poised for a Comeback?" *Science* 309, no. 5738 (2005): 1168–1169. 10.1126/science.309.5738.1168.

Milly, P.C.D., Julio Betancourt, Malin Falkenmark, Robert M. Hirsch, Zbigniew Z. Kundzewicz, Dennis P. Lettenmaier, and Ronald J. Stouffer. 2008. "Stationarity is Dead: Whither Water Management?" *Science* 319, no. 5863: 573–574. 10.1126/science.1151915.

Morrison, Robert. 2001. "Global Nuclear Markets in the Context of Climate Change and Sustainable Development." In *Canadian Nuclear Energy Policy: Changing Ideas, Institutions, and Interests*, edited by G. Bruce Doern, Arlsan Dorman, and Robert W. Morrison, 34–51. Toronto: University of Toronto Press. http://www.jstor.org/stable/10.3138/9781442672260.6.

National Academies of Sciences, Engineering, and Medicine. 2016. *Attribution of Extreme Weather Events in the Context of Climate Change*. Washington, DC: The National Academies Press. 10.17226/21852.

Noriyoshi, Takemura. 2020. "The Hottest Chaos of Cobalt Mining in the Democratic Republic of Congo: 'Triadic Abyss' of Human Rights Abuses, Environmental

Pollution, and Illegal Global Trade." *Toin University of Yokohama Research Bulletin* 42: 35–42. 10.50937/00000361.

Norwegian Church Aid and Catholic Commission for Justice and Peace (Malawi). 2013. *Malawi's Mining Opportunity: Increasing Revenues, Improving Legislation.* Lilongwe, Malawi: Norwegian Church Aid. https://curtisresearch.org/malawis-mining-opportunity-increasing-revenues-improving-legislation/.

NTI Nuclear Security Index. n.d. Accessed June8, 2021. https://www.ntiindex.org/.

Nyirenda, William. 2014. "Paladin Lies to Malawi Government on its Kayelekera Uranium Mine." Environmental Justice Organisations, Liabilities and Trade, April 2, 2014. http://www.ejolt.org/2014/04/paladin-lies-to-malawi-government-on-its-kayelekera-uranium-mine/.

Pan-Amazonian Ecclesial Network (Red Eclesial Panamazónica). 2019. "Situación de urgencia: amenaza de muerte a líderes en Colombia, llamado internacional a proteger la vida de los defensores." January 21, 2019. http://redamazonica.org/2019/01/situacion-de-urgencia-amenaza-de-muerte-a-lideres-en-colombia-llamado-internacional-a-proteger-la-vida-de-los-defensores/.

Percebois, Jacques. 2003. "The Peaceful Uses of Nuclear Energy: Technologies of the Front and Back-Ends of the Fuel Cycle." *Energy Policy* 31, no. 2: 101–108. 10.1016/S0301-4215(02)00035-6.

Powers, Gerard F. 2015. "From Nuclear Deterrence to Disarmament: Evolving Catholic Perspectives." *Arms Control Today* 45. https://www.armscontrol.org/act/2015-05/features/nuclear-deterrence-disarmament-evolving-catholic-perspectives.

Powers, Gerard F. 2017. "Toward an Integral Catholic Peacebuilding." *Journal of Social Encounters* 1: 1–13. https://digitalcommons.csbsju.edu/social_encounters/vol1/iss1/2/.

Princen, Thomas. 2002. "Distancing: Consumption and the Severing of Feedback." In *Confronting Consumption*, edited byThomas Princen, Michael Maniates, and Ken Conca, 103–132. Cambridge, MA: MIT Press. 10.7551/mitpress/2097.001.0001.

Rénique, Gerardo. 2006. "Strategic Challenges for Latin America's Anti-Neoliberal Insurgency." In *Dispatches from Latin America: On the Frontlines against Neoliberalism*, edited by Teo Ballvé and Vijay Prashad, 35–46. Cambridge, MA: South End Press.

Richter, Burton. 2006. "Between 2 Devils." *IAEA Bulletin* 47, no. 2: 14–17. https://www.iaea.org/sites/default/files/publications/magazines/bulletin/bull47-2/47204791417.pdf.

Ripley, Earle A., E. Robert Redmann, Adèle A. Crowder, Tara C. Ariano, Catherine A. Corrigan, Robert J. Farmer, L. Moira Jackson. 1996. *Environmental Effects of Mining*. Boca Raton, FL: Routledge. 10.1201/9780203757147.

Roman Catholic Bishops of New Mexico. 1998. "Partnership for the Future." https://archdiosf.org/partnership-for-the-future.

Sampat, Payal. 2003. "Scrapping Mining Dependence." In *State of the World 2003: A Worldwatch Institute Report on Progress Towards a Sustainable Society*, edited by Lisa Starke, 110–129. London: Routledge. 10.4324/9781849776257.

Strongman, John. 1994. "Strategies to Attract New Investment for African Mining." https://documents1.worldbank.org/curated/en/349851468767422230/pdf/multi-page.pdf.

Thomas, Patricia A., and Thomas E. Gates. 1999. "Radionuclides in the Lichen-Caribou-Human Food Chain Near Uranium Mining Operations in Northern Saskatchewan, Canada." *Environmental Health Perspectives* 107, no. 7: 527–537. 10.1289/ehp.99107527.

Wallace-Wells, David. 2019. *The Uninhabitable Earth: Life After Warming*. New York: Tim Duggan Books.

World Bank. 1992. *Strategy for African Mining*. World Bank Technical Paper Number 181, Africa Technical Department Series. Washington, DC: World Bank. https://documents1.worldbank.org/curated/en/722101468204567891/pdf/multi-page.pdf.

Concluding reflections

Laurie Johnston

God could be considered the first miner. In Genesis, God draws dirt out of the earth, shapes it, and transforms it into human persons. Created in the image of God, we are invited to share in that creative work, and so we too turn the earth for the raw materials it requires. It is impossible to conceive of the human vocation on earth without mining, for "everything that comes from man throughout the whole process of economic production ... presupposes these riches and resources of the visible world, riches and resources that man finds and does not create" (John Paul II 1981, §12). The precious minerals that we find in the earth are part of God's generosity and the goodness of creation, as Cardinal Turkson points out in the first chapter of this volume. Yet, like all of creation, they are in "bondage to decay" as a result of sin (Rom. 8:20–21). As the chapters throughout this volume reveal, mining projects have often revealed the extent of human sin, violence, and injustice. Mining presents a serious challenge to Christians' witness to God's redemption of *all* creation and our vocation to become artisans of an ecological just peace.

Examining the roots of a complex problem

The connection between violent conflict and mining has complex causes, contours, and possible solutions. As the case studies from Section 1 reveal, there are many sites of extraction where the "cry of the poor" and the "cry of the earth" ring out in unison. *Querida Amazonia* summarizes:

> The businesses, national or international, which harm the Amazon and fail to respect the right of the original peoples to the land and its boundaries, and to self-determination and prior consent, should be called for what they are: injustice and crime. When certain businesses out for quick profit appropriate lands and end up privatizing even potable water, or when local authorities give free access to the timber companies, mining or oil projects, and other businesses that raze the forests and pollute the environment, economic relationships are unduly altered and become an instrument of death. They frequently resort to

DOI: 10.4324/9781003094272-102

> utterly unethical means such as penalizing protests and even taking the lives of indigenous peoples who oppose projects, intentionally setting forest fires, and suborning politicians and the indigenous people themselves. All this accompanied by grave violations of human rights and new forms of slavery affecting women in particular... (Francis 2020, §14)

While this refers specifically to the Amazon, mining and conflict are often "locked in an escalating spiral," as Karl Gaspar puts it. Similar attitudes, theologies, and value systems seem to underly both violence against the earth and violence against people, shaping the social sin of "extractivism":

> Extractivism is a short-sighted model of development seen around the world that exploits natural resources on a massive scale, creating significant economic profits for the powerful few in the short term, but too often resulting in minimal benefits for the communities where these resources are found. (Inter-Religious Working Group on Extractivism 2019)

When the profit motive is supreme, it is easy to overlook the fact that resources are limited and will eventually be exhausted. Holden and Montevecchio write, "What the extractivist paradigm never contemplates is a cessation of extraction; the ideas of removing less from the earth ... or else changing consumption patterns altogether are never considered."

Extractivism is a corollary to what Pope Francis has called "throwaway culture." The term powerfully expresses the interconnected evils evident in the world today, reflected in human attitudes and behavior but also in economic and political structures. Like throwaway culture, extractivism is rooted in a disavowal of connectedness. It seeks only how to *use* resources, and often other people, whether through slavery, colonial domination, or other forms of dehumanization. For centuries, extraction has been a central motivation for colonial and imperial expansion and has been associated with many forms of violence, particularly against the bodies of women and indigenous peoples. Thus, the problems that are described in this volume are neither new nor particularly surprising.

Mining plays an enormous role in human flourishing today, too. As William George (2019, x) has said, "Everything begins with mining." Virtually anything we touch comes from extraction. Moreover, the transition to renewable energy requires minerals to manufacture solar panels, batteries, and other green technologies; the fate of the planet depends upon the continuation of mining. Yet its fate (and the fates of many humans) are also threatened by what is fundamentally an unsustainable enterprise—minerals do not renew themselves. This is why some of the authors in this volume question whether mining can ever be done ethically. Raymond Offenheiser describes some of the significant improvement major mining companies have made, but there is no doubt that extraction always comes with moral perils and environmental risks. In some places, those risks are simply too great; Andrés McKinley explains how Salvadorans came to conclude that their country was one of those places. Still, mining will continue, which is why we must, as Tobias Winright says, try to *limit* the evils associated with it. This requires a holistic response—a truly *integral* approach to peacebuilding.

Challenges and opportunities

When it comes to engaging this nexus of mining, violence, and ecological destruction, what do Catholic peacebuilders have to bring to the table? There are a number of important resources and opportunities that emerge from the Catholic tradition, and some significant challenges.

Peace as the fruit of right relationships

Catholic social teaching frequently reminds us that peace is the fruit of right relationships. This volume has illustrated breakdowns in many types of relationship, and peacebuilding in the context of mining requires attention to each of them. Authors in this volume have addressed the complex triad of mining companies, national governments, and local communities where mining takes place. On the one hand, it is clear that communities need practical and pastoral assistance in holding their governments accountable. When governments receive significant portions of their income from mining, they are more likely to look away while local communities are displaced, face ecological damage, or experience human rights violations at the hands of private security forces (or even national police). On the other hand, even when governments do seek to protect the interests of their citizens, they are at a significant disadvantage; large mining companies have technical, informational, financial, and legal resources that outmatch most governments in the Global South. Many countries have faced lawsuits from deep-pocketed mining companies seeking to preserve their exploitation rights. Finally, even mining companies seeking to do the right thing may encounter obstacles from government officials or local militias who seek to extort them, or competition from other mining companies who are less scrupulous. In this context, peacebuilders who seek to promote just relationships must find ways to empower local communities to hold governments accountable, promote good governance for the common good, and continue to push both large and smaller mining companies to conduct themselves in more and more ethical and sustainable ways.

Another important triad is the relationship between governments, citizens, and mining companies in the countries where these mining companies are based. As Vincent Miller points out, citizens of wealthy countries are largely in the dark about the sourcing of consumer products; they enjoy many of the benefits of extractive industries while bearing few of the costs. Governments of wealthy countries are often captive to mining interests, too; William Holden's account of the Canadian ambassador in Guatemala intervening on behalf of Canadian mining interests is a telling one. A key task, then, is finding ways to make citizens of wealthy countries more aware of the human and ecological consequences of their consumer behavior—not only to incentivize changes in this behavior, but so that they will push their governments to establish laws and policies that will bring about shifts in the global economic system, such as Dodd-Frank and some European countries' emerging laws about responsible multi-national corporations. These wealthy countries also play an outsize role at the WTO, at the UN, and in governing the World Bank, all of which play important, but distinct, roles in shaping the global economic system and the mining industry.

The Catholic Church has a powerful ability to bear witness to what is happening in many different places around the world. Several times in the past decade, major mining executives have attended meetings at the Vatican—in part because they realize that the church is present in nearly all the countries in which they operate. This broad reach is important precisely because mining takes place in remote areas where oversight is difficult and government authorities may be quite limited. Human rights violations can take place with relative impunity. The coronavirus pandemic has made oversight and monitoring even more difficult. And while major mining companies have made progress towards more ethical practices, small and mid-range mining operations frequently remain opaque. Thus, the Catholic Church's ability to monitor mining projects and advocate for local communities will continue to be important. Earlier chapters in this volume testify to the variety of ways the church has played this role, from assessing contracts in the Democratic Republic of the Congo (DRC), to conducting pastoral surveys in Colombia, to issuing pastoral letters like the Philippine bishops' "What is Happening to Our Beautiful Land?"

Oversight and monitoring are greatly aided by transparency. As the Extractive Industries Transparency Initiative makes more information about mining operations and contracts accessible, it is important for there to be an informed public that can consume this information and compare it with realities on the ground. The church's ability to bear witness is thus an important adjunct to the ongoing push for greater transparency. However, this also poses a challenge to Catholic leaders: much like mining companies, the Catholic Church does not have a strong tradition of institutional transparency. Furthermore, the church can itself be vulnerable to the potentially corrupting influence of donations from those with vested interests in extractive industries. Catholic University of America has been criticized for accepting millions of dollars from the Koch Foundation, given that the Koch brothers are oil magnates who have sought to protect extractive industries by funding climate skeptics and lobbying against policies that aim to reduce global warming (McElwee 2013).

In the Philippines, Father Benvenido Miguel has criticized Catholic leaders for soliciting donations from mining companies, and even accepting cars and hotel vouchers: "Asking favors or accepting gifts from companies 'creates a perception that they can buy and silence the Church' on issues of environmental protection" (Saludes 2020, n.p.). Such potentially corrupting forces exist in many Catholic contexts; the best protection against them is ensuring a strong commitment to internal *and* external transparency as a key component of the church's witness. Right relationships among all in society, not merely cozy relationships between elites, are the key to sustainable peace.

Becoming truly "catholic"

In addition to bearing witness to what is happening in local communities around the world, a second important role for the church is creating connections among Catholics and between Catholics and other groups. According to McKinley, Catholic leaders facilitated a visit from a Philippine governor that was instrumental in convincing communities and politicians in El Salvador to be skeptical of the role mining might play in their country. The testimony of a Congolese bishop before the US Congress, supported by the US Catholic bishops, helped garner support for Dodd-Frank. Ecclesial networks can help connect those who bear the impact of mining in different countries in order to, as Offenheiser puts it, "link grievances at the community level to patterns of abuse at the national and global level and then turn the observation of these patterns into very specific policy recommendations to the industry." Miller calls especially for linking those impacted by mining with consumers and political leaders in wealthy countries that benefit from mining, to create "deeper relationships of solidarity." This book is itself an attempt to contribute to that effort.

Mining also offers many opportunities for collaboration with groups outside the church. In fact, the Catholic Church is somewhat late to the party; Oxfam and other organizations have played major roles in pushing for ethics and sustainability in the extractive industries, and the church can benefit from their expertise. Partnerships with other religious communities are also vital, particularly in places like the Philippines and Mozambique where there are overlaps between religious conflicts and mining-related conflicts. Elias Opongo has argued that these sorts of broad coalitions are key for church action to be effective, especially at the level of governance and policy. As Katherine Marshall points out, there is often no "accepted 'place' for religious engagement in many of the international debates that affect extractive industry policies or in efforts to address problems that arise." Yet by forming alliances, religious actors can have a better chance of gaining a place at the table.

Building these intra-ecclesial and extra-ecclesial connections raises ecclesiological questions, however. First, several of the authors in this volume call for a greater role for the laity, given that technical expertise is vital in engaging with mining. At the same time, it is clear that clergy, especially bishops, have a high level of visibility and moral authority in countries like the DRC, and therefore have a very important role to play. In other contexts, Catholic laypeople, religious orders, universities, and civil society organizations have taken the lead. Of course, there may also be significant overlap between the efforts of laity, religious, and clergy; DHUMA, as described by José Bayardo Chata Pacoricona, offers an interesting example of an organization that began from a collaboration between bishops, religious congregations, and lay parish workers, and now operates as an independent civil society organization while maintaining its Catholic identity.

Perhaps the greatest challenge for Catholic engagement in mining is finding ways to truly incorporate women's voices and leadership. The leadership of the mining industry and the leadership of the institutional church are both male-dominated. Yet the greatest negative impacts of mining accrue to women. Mining is nearly always accompanied by sexual exploitation of various kinds, and the conflicts that surround mining involve gender-based violence on a massive scale. The environmental effects of mining also burden women in particular: in many contexts, it is they who are primarily responsible for farming and for supplying water to their families. Yet their concerns are frequently overlooked; even when community consent processes do take place, they often neglect women's participation and fail to account for social structures and mores that inhibit that participation.

At the same time, women are frequent activists for justice in and beyond mining communities; their approaches have much to teach others who are seeking to promote an ecological just peace. For instance, indigenous women in Guatemala have called attention to the importance of thinking about the long-term. One activist explained, "Women take the time to think: what will happen tomorrow? Or what will I do tomorrow? Men do not think this way" (Deonandan, Tatham, and Field 2017, 409). Other women have found creative ways to push back against the exploitation of their bodies that frequently accompanies extractive industries. In South Africa, women seeking employment in the mines were angered by incidents of men demanding sex in exchange for a job. To push back, women in Mpumalanga adopted the slogan, "My body is not my CV" (WoMin African Alliance 2020). There are many other cases; a major weakness of this present volume is that it does not give sufficient attention to these voices and examples. While Catholic engagement—including recent Vatican meetings—has frequently included women religious, real gender justice requires much more. Environmental activists have coined the term "greenwashing" to refer to shallow or disingenuous attempts to appear environmentally responsible. The term "pinkwashing" can refer to analogous attempts to include women's voices while still excluding them from power structures and from determining the overall agenda. Both the church and mining companies must pay particular attention, then, to the critiques coming from organizations like WoMin in Africa.

Seeking the common good

The violent conflicts that often emerge around mining illustrate what happens when certain goods or resources are pursued without attention to a larger common good. Authors in this volume have discussed the common good, the ecological common good, and even the cosmic common good. However, when there are conflicting interests in society, determining how best to promote the common good is difficult. When mining executives arrived at the Vatican for a day of reflection in 2013, Vatican Secretary of State Tarcisio Bertone said: "The great challenge of business leaders is to create a harmony of interests, involving investors, managers, workers, their families, the future of their children, the preservation of the environment on both a regional and international scale, and a contribution to world peace" (Vatican Radio 2013, n.p.). However, it is a real question whether such a "harmony of interests" is actually possible. Kenneth Himes has noted that a "failure to acknowledge the deeply conflictual nature of human reality has permitted Catholic social teaching to remain underdeveloped in strategies of conflict resolution" (Himes 2010, 282–3). Human sinfulness and the conflictual nature of reality mean that tensions around mineral resources are *to be expected*. There are situations where there is no "win-win" outcome, and Catholic social teaching has not always accounted for these well. If a community rejects a mining project, investors lose money and resources for potentially good ends are lost; it is not surprising that mining companies remain reluctant to fully embrace the idea that they must respect community consent as a binding principle, as it might mean abandoning promising projects on which millions of dollars have already been spent.

Still, there *are* areas of common interest between extractive industries and local communities, and the Catholic Church can play a role in finding these, for the sake of the common good. Many of the largest mining companies have come to understand that upfront investment in good relationships with local communities is in their best interests, particularly because this may prevent conflict, fines, and even premature mine closures. Some mining companies have been in favor of greater financial transparency; they want to disarm opposition by proving that they are paying their fair share and ensuring that their payments are not lost to corruption. Another possible area of common interest is infrastructure development. Mining endeavors frequently require companies to develop electric, water, and transportation infrastructure; it is often possible to do this in a way that will offer lasting benefits to local communities. Albino Barrera describes the problem of "export enclaves," and partnering with local communities to develop this infrastructure could be a good way to break away from that enclave model. To capitalize upon this, there is a need for community leaders who can liaise with mining companies and government officials in order to find these areas of common interest, and local church leaders and women religious can and do play that role.

Prophecy vs. collaboration

One challenge that Catholic leaders face is this: when should they take a stance of collaboration, and when should they focus on holding companies accountable or opposing their work outright? Given the devastating effects of mining on people and the environment, a prophetic critique seems like the appropriate response in many circumstances. Daniel Finn (2017, n.p.) agrees, but also calls for a nuanced understanding of the church's prophetic role:

> [T]he way to get abusive mining companies to shift to the new paradigm is to make it less arduous for the responsible firms than for the irresponsible ones. The world needs a prophecy that remains open to the possibility of authentic transformation. And where transformation does indeed occur, prophets need to shift their methods toward civil engagement for a more just and sustainable world. Finn 2017

It is difficult to find a mode of public engagement that holds corporations and governments accountable while resisting the temptation to demonize the people within these institutions. Catholics seeking to engage mining companies must strike a difficult balance: how do we recognize moral progress but maintain a healthy skepticism about an industry where the human risks, environmental risks, and moral hazards are enormous?

Questions around incremental moral improvement are a challenge in many areas of ethical reflection. Roger Burggraeve (1994) has sought to address them through an "ethics of growth." This approach is rooted in the idea that even when individuals are in clear violation of moral norms, that does not mean the task of providing moral guidance ends. Rather, the church should examine how to help all persons continue to grow in the moral life. Such an approach can provide a way for Catholic activists who oppose most mining in principle to nevertheless seek ways to make sure it is conducted less harmfully. Similarly, an "ethics of growth" can be a way to describe how Catholic peacebuilders engage violent actors in moving towards peace, while not overlooking the real evils for which these actors may be responsible. Such an approach testifies to faith in the redeeming power of God that may emerge in any situation, bringing the common good to fruition despite obstacles.

The longue durée of war and mining

A key strength of Catholic peacebuilding has been its focus on a broad time horizon. Scott Appleby (2010, 3) has defined peacebuilding as "conflict transformation that strives to comprehend the *longue durée* of a conflict." Political attempts at peacemaking fail because of their short time horizon; a Catholic vision of what peace requires is far more holistic and requires us to take a long view and strive for something more sustainable. Creating a just peace requires attention to root causes that may be long in the past, as well as a commitment to persevere in practices of peacebuilding that may take generations to bear fruit. Recently, theorists of *jus post bellum* have called for greater attention to the long-term consequences of war. What is wrong about war is not just the destruction that happens during war, but the aftereffects for both people and the earth: landmines, depleted uranium, unexploded ordnance, napalm-poisoned soil. Traumatized communities may take generations to be reconciled.

The ethics of mining also has a challenge with time—and we have seen already that this is an area of particular critique from women activists. While some mining companies have embraced the idea of planning for a mine's ultimate closure from the very beginning, others have been slow to embrace the idea of responsibility for the entire lifecycle of a mine. For instance, when a large mining company has exhausted the high-quality ores in a mine, they may sell it off to a smaller mining company that may choose to mine lower-quality ores (which is likely to be less efficient and more disruptive) and then hurriedly close the mine without doing the appropriate cleanup or without ensuring the stability of tailings dams. Even when a mining project begins responsibly and obtains community consent, mining personnel may change over time, and important policies and practices may slide into oblivion. As Holden and Montevecchio point out in their chapter, the geologic timescale of these undertakings is vast—particularly when it comes to radioactive ores—and so *perpetual* attention is required.

How, then, can we cultivate forms of social virtue that endure for the long term? Law can help to serve this purpose, of course—and so it is important, as Offenheiser and Douglass Cassel urge, to continue the push for hard law that protects the earth and empowers communities (not just voluntary principles, as many mining companies would prefer). But this is also a task for churches, as spaces of formation, education, and memory. Pope Francis' call for intergenerational solidarity in *Laudato Si'* is an important way to nurture our attention to a longer time frame. Churches can do more to link liturgical practices to advocacy, perhaps by calling attention to the ways that mining often threatens the water that we use for baptism and the farmland on which wheat is grown for the Eucharist. Also, Sandra Polanía-Reyes and Héctor Fabio Henao describe how the Church in Colombia has focused intently on pastoral work to help communities become more attuned to a theology of creation.

The doctrine of creation is vitally important as a justification for care for the earth. However, Catholics also must understand environmental activism from the broader perspective of eschatological time. As the Scheids point out, the Christian mandate to respect the earth is also rooted in a vision of its *destiny*. In *Laudato Si'*, Pope Francis (2015, §243) calls us to care for our earthly home "which has been entrusted to us, knowing that all the good which exists here will be taken up into the heavenly feast." What we do to nature *matters*, because the earth will share with us in the final salvation that is coming. In his epistle to the Romans, Paul tells us that "All of creation is groaning in anticipation ..." of redemption, and that includes the rocks and ores in the ground. Of course, the Scheids point out that "bearing in mind our shared eschatological future or the full cosmic import of our choices will not fully or even perhaps partially determine concrete practical political solutions toward peacemaking and social justice." Still, it is possible that we need to abandon a key idea that moral theologians have used for centuries—"the universal destination of created goods." The idea that God destined creation to be used by all of humanity is a powerful way to limit excessive wealth or the hoarding of mineral resources by just a few. Pope Francis uses this phrase in *Laudato Si'*, and it would be excellent if humans could even come close to the sharing to which that phrase calls us. But in the Anthropocene, we must no longer speak about the earth as "destined" for human consumption. Yes, we can and should use the goods of the earth, in a way that is proportional, as Winright suggests. But the earth has its own destiny and dignity, even, perhaps, its own rights. Better, then, to think of *friendship* with the earth as we move together towards our shared redemption. By merging the idea of integral ecology with Pope Francis' (2020) call for social friendship in *Fratelli Tutti*, we can cultivate a new practice of ecological friendship.

When there are broken relationships, however, moving towards friendship with the earth requires reconciliation. Reconciliation is a major theme of Catholic peacebuilding; lasting peace is impossible without it, as many conflicts around the world have shown. But how do we reconcile with the earth that we have damaged? Truth-telling and acknowledgment of environmental damage for which we are responsible are certainly important, as are reparations that seek to restore ecosystems. Memorialization is also vital, partly for the sake of future generations who might forget just how much damage can be done to both humans and the earth by unbridled extractivism. In particular, it is important to acknowledge the damage that is done when people are displaced from their land—a particular kind of broken relationship that is common in the context of mining. As Clemens Sedmak writes, "When rural or indigenous communities face losses from a mining project that is supposedly good for economic development, they may officially gain access to higher 'living standards,' but they lose identity-conferring aspects of their lives." Loss of identity is a key driver and consequence of many violent conflicts around the world. Finding ways to acknowledge this kind of deep loss in the liturgical life of the church is key for building peace and bringing healing in human communities and with the earth as well—even if the true peace we long for will arrive only at the eschaton.

A peacebuilding lens as a new way of seeing

Pope Francis reminds us that there is no single solution to the interconnected challenges we face today. In fact, technological specialization often makes it "difficult to see the larger picture" (Francis 2015, §110). Instead, he says, "There needs to be a distinctive way of looking at things, a way of thinking, policies, an educational programme, a lifestyle and a spirituality which together generate resistance to the assault of the technocratic paradigm" (§111). The chapters in this volume are an invitation to look at mining-related conflicts in a more holistic way. The task also invites us to see our own tradition and Scriptures in new ways. Perhaps Isaiah's vision of peace can also be a reminder about the careful repurposing of the earth's minerals: "they shall beat their swords into ploughshares, and their spears into pruning-hooks" (Is. 2:4). An ecological just peace requires us to *see* the intrinsic dignity of one another and of the earth itself, in light of our shared destiny. Such a transformed way of looking at the world ultimately requires conversion from all of us.

References

Appleby, R. Scott. 2010. "Peacebuilding and Catholicism: Affinities, Convergences, Possibilities." In *Peacebuilding: Catholic Theology, Ethics, and Praxis*, edited by Robert J. Schreiter, R. Scott Appleby, and Gerard F. Powers, 3–22. Maryknoll, NY: Orbis.

Burggraeve, Roger. 1994. "Une éthique de miséricorde." *Lumen Vitae* 49: 281–296.

Deonandan, Kalowatie, Rebecca Tatham, and Brennan Field. 2017. "Indigenous Women's Anti-Mining Activism: A Gendered Analysis of the El Estor Struggle in Guatemala." *Gender & Development* 25, no. 3: 405–419. 10.1080/13552074.2017.1379779.

Finn, Daniel. 2017. "Recycling isn't enough: A Defense of Responsible Mining." *Commonweal*, May 2, 2017. https://www.commonwealmagazine.org/recycling-isn't-enough.

Francis. 2015. *Laudato Si'*. http://www.vatican.va/content/francesco/en/encyclicals/documents/papa-francesco_20150524_enciclica-laudato-si.html.

Francis. 2020. *Querida Amazonia*. http://www.vatican.va/content/francesco/en/apost_exhortations/documents/papa-francesco_esortazione-ap_20200202_querida-amazonia.html.

Francis. (2020). *Fratelli Tutti*. https://www.vatican.va/content/francesco/en/encyclicals/documents/papa-francesco_20201003_enciclica-fratelli-tutti.html.

George, William P. 2019. *Mining Morality: Prospecting for Ethics in a Wounded World*. Lanham, MD: Lexington Books.

Inter-Religious Working Group on Extractivism. 2019. "A Just Response to Extractivism." https://justresponse.faith/node/3.

John Paul II. 1981. *Laborem Exercens*. https://www.vatican.va/content/john-paul-ii/en/encyclicals/documents/hf_jp-ii_enc_14091981_laborem-exercens.html.

McElwee, Joshua J. 2013. "Republican Donors' Gift to Catholic University Comes Under Fire." *National Catholic Reporter*, December 16, 2013. https://www.ncronline.org/news/politics/republican-donors-gift-catholic-university-comes-under-fire.

Saludes, Mark. 2020. "Some Philippine Catholic Church Institutions Accept 'Dirty Donations.'" *Licas News*, September 29, 2020. https://www.licas.news/2020/09/29/some-philippine-catholic-church-institutions-accept-dirty-donations/.

Himes, Kenneth. 2010. "Peacebuilding and Catholic Social Teaching." In *Peacebuilding: Catholic Theology, Ethics, and Praxis*, edited by Robert J. Schreiter, R. Scott Appleby, and Gerard F. Powers, 265–299. Maryknoll, NY: Orbis.

Vatican Radio. 2013. "Pope Hopes Day of Reflection on Mining Ethics Will Benefit Industry, Workers." http://www.archivioradiovaticana.va/storico/2013/09/09/pope_hopes_day_of_reflection_on_mining_ethics_will_benefit_industry,/en1-726839.

WoMin African Alliance. 2020. "Women Stand Their Ground Against Big Coal." https://womin.africa/download/women-stand-their-ground-against-big-coal-final/.

Index